W9-BZO-993

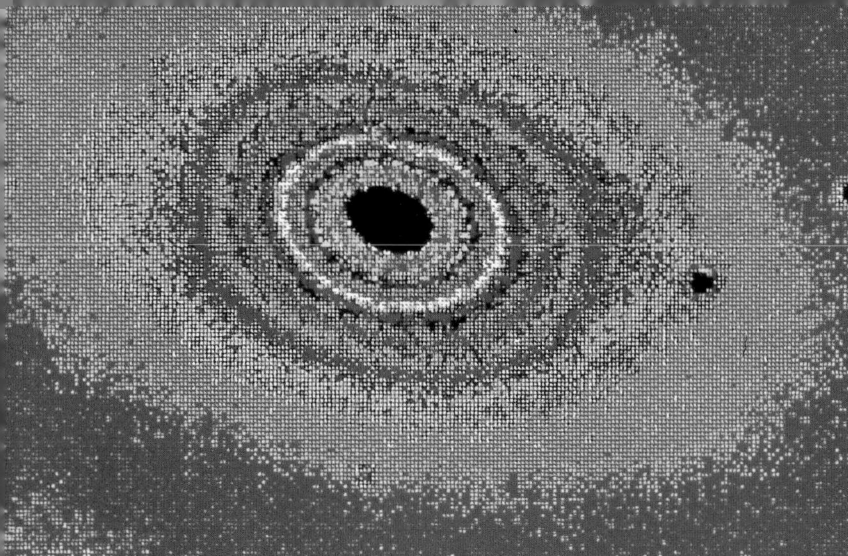

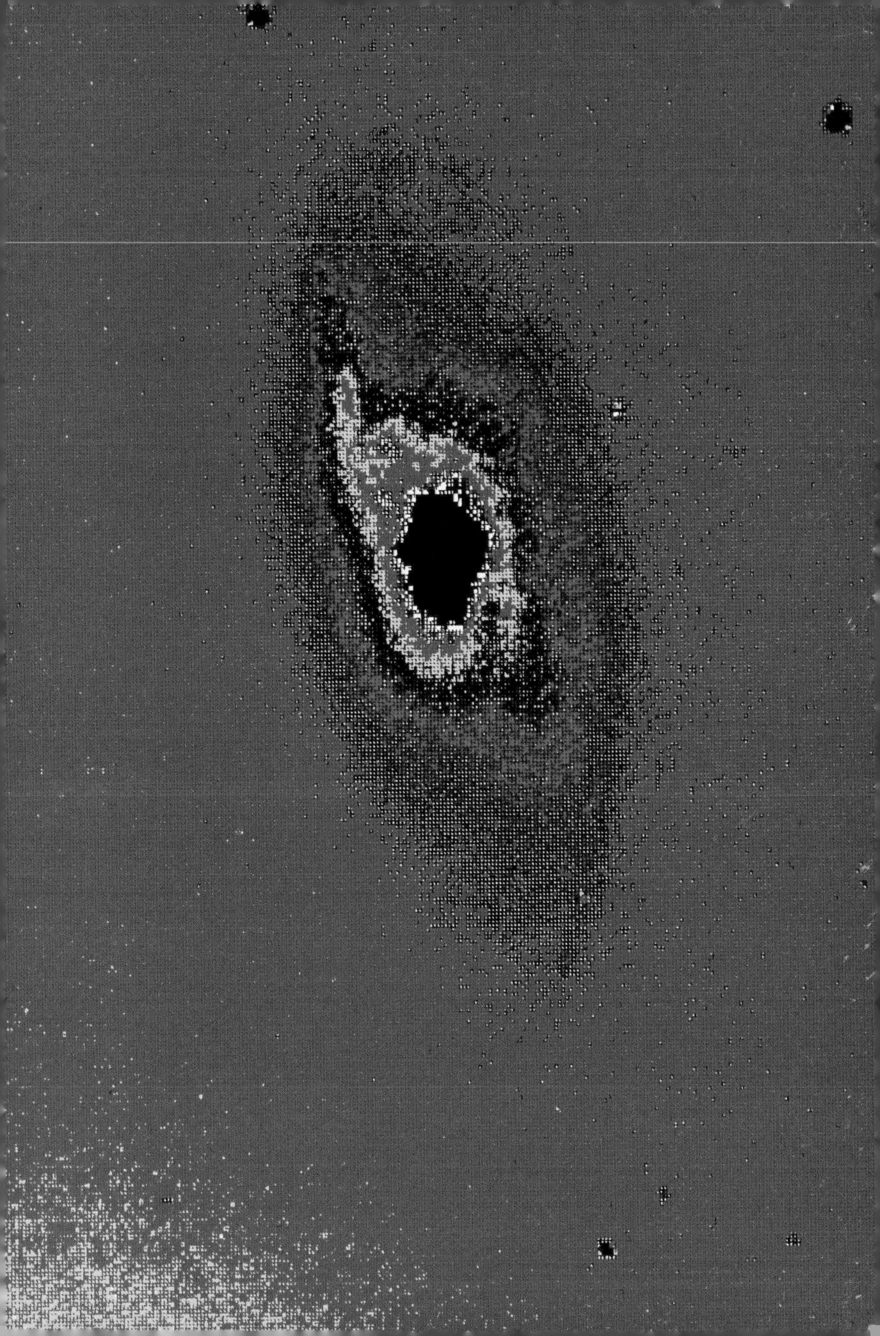

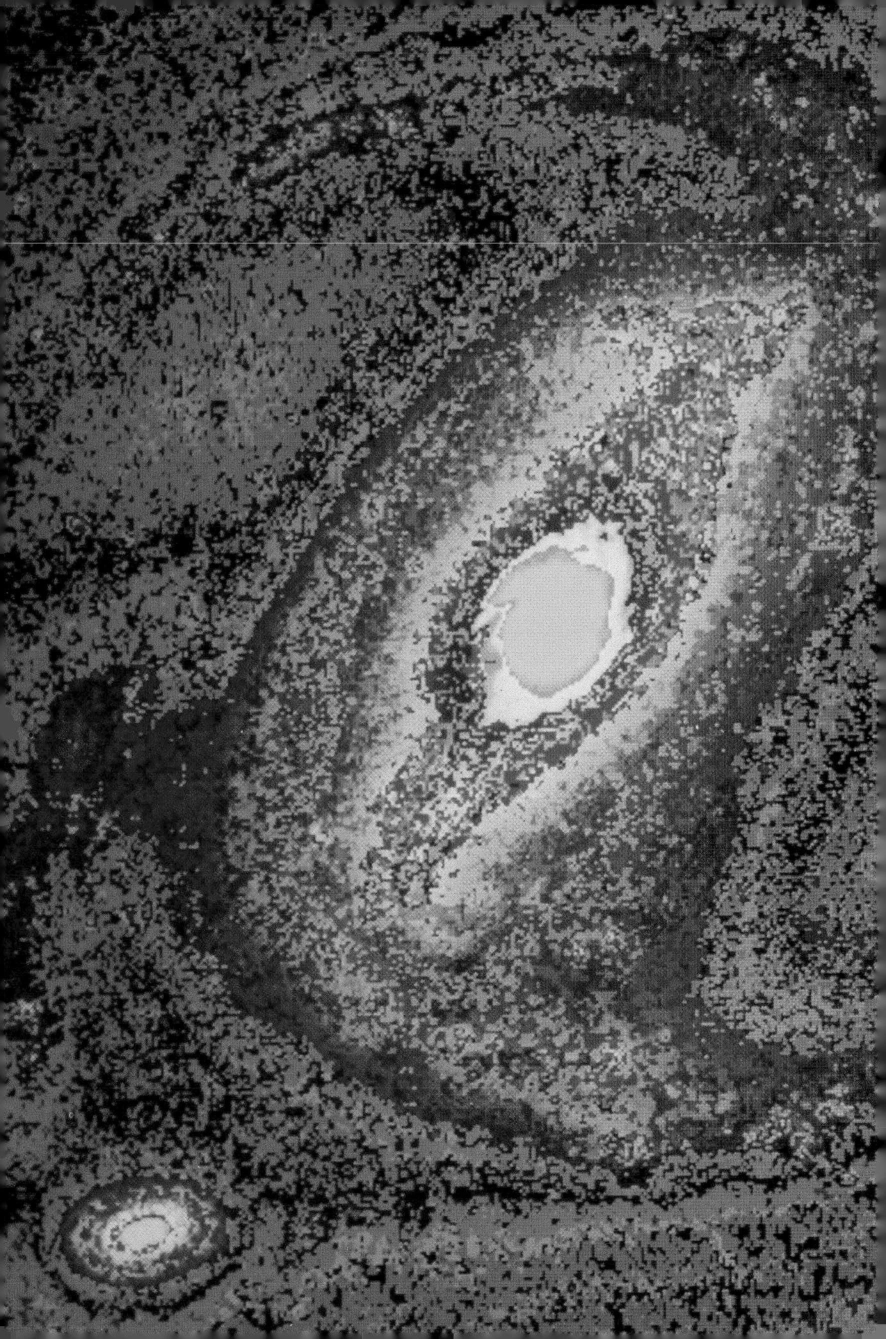

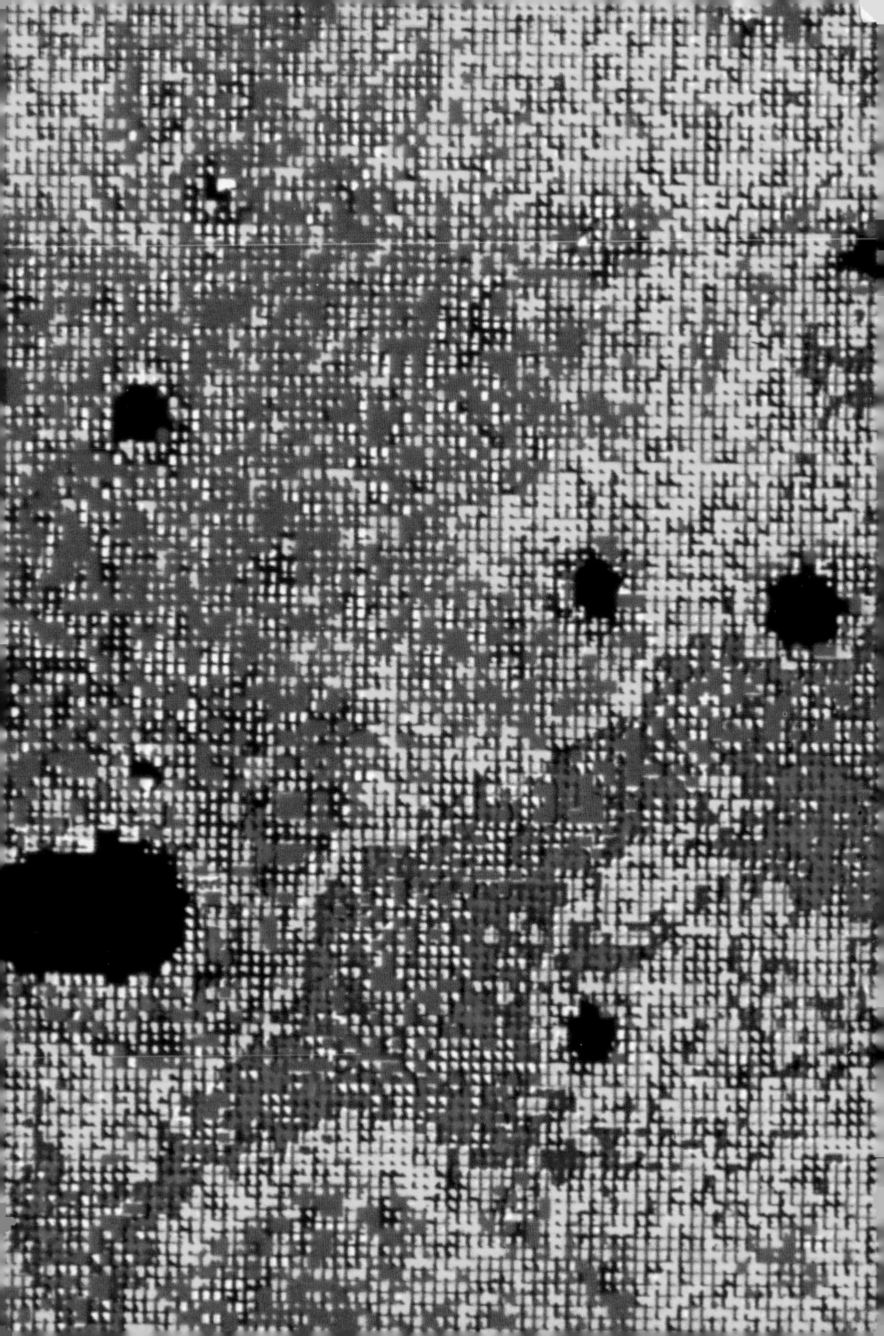

CONTENTS

THE BIRTH OF THE UNIVERSE
THE BIG BANG AND BEYOND

Trinh Xuan Thuan

DISCOVERIES
HARRY N. ABRAMS, INC., PUBLISHERS

Humankind has always sought to puzzle out the secret melody of the universe, to order into a logically consistent totality phenomena as seemingly disparate as the light of day and the dark of night, the mauve glimmerings of sunrise and ruddy glow of sunset, and the swath of pale white arching across the vault of heaven on clear summer nights.

CHAPTER I

WORLD VIEWS

Humans of all eras and cultures have projected their longings and musings onto the heavens. The Egyptians pictured the sky as the body of the fair goddess Nut (left). By the Middle Ages, people realized that the earth was round but clung to the belief that the positions of planets relative to the constellations of the zodiac (right) could decide the fate of individuals and nations.

The Egyptian sun-god Ra, shown in this painted wooden stela (2480–2350 BC), sheds warmth and energy on a female votary.

When Spirits and Deities Ruled the Universe

From the very beginning, human beings have attempted to dispel their mortal dread of infinite space by structuring the world around them in ways that make it seem a more familiar place.

The universe pre-Neanderthal humans lived in hundreds of thousands of years ago was a world of spirits: the sun spirit by day, the moon and star spirits by night, the tree spirit that brought forth fruit, the stone spirit that tripped them up. Theirs was a reassuring, familiar world tailored to the human point of view.

Mythic worlds ruled by deities started emerging some 10,000 years ago. The exploits of gods and goddesses —their dalliances and couplings, their enmity and strife—were believed responsible for all natural phenomena, including the creation of the universe.

Women, bearers of children, inspired many a creation myth. Five thousand years ago Babylonians believed that Tiamat, the primordial woman, and Apsu, god of the underworld ocean, begat the sky-god Anou and that Anou and Tiamat then begat the water-god Ea.

The primeval ocean was also the source of life in ancient Egyptian mythology. It was the dwelling place of the first being, Atum, repository of all existence and later identified with the sun-god Ra. The earth-god Geb, a flat disk surrounded by mountains, floated on the primeval ocean. Propped up by the air-god Shu, the body of the fair goddess Nut formed the vault of the sky and sparkled with the jewels mortals saw as planets and stars. Ra completed his daily circuit across the heavens by sailing across Nut's back on a barge and plied the underworld waters by night to return to his starting point in the east.

In Chinese cosmology*, there were no personified gods. The universe came into being through the combination and interaction of two opposite forces, yin and yang. The sky was yang, the strong, creative, masculine principle. The earth was yin, the maternal, feminine principle. Yin and yang alternated in a never-ending cycle: The hot, dry light of the sun (yang) gave way to the dark, cold, wet light of the moon (yin).

In Indian mythology, Shiva represents the everlasting energy of the cosmos. He is shown here as the four-handed Lord of the Dance of Creation, surrounded by an aureole of fire emanating from a lotus, emblem of enlightenment. The prostrate form on which he is dancing represents ignorance. The drum in his upper left hand symbolizes the music of creation, while the tongue of flame in his upper right hand portends the future destruction of the universe. According to Indian myth, the universe goes through cycles of life and death spanning billions of years—a concept that brings to mind certain aspects of modern cosmology. Everyone knows that the universe is expanding, but its future course is anything but certain. If the average density of matter in the universe is sufficiently high, gravity will eventually overcome expansion and the universe will one day collapse on itself and end in an infinitely hot and dense "Big Crunch." The positions of Shiva's two front hands symbolize the eternal balance of life and death.

* See glossary starting on p.150 for an explanation of this and other terms.

Although technologically far more advanced than medieval Europe, China—birthplace of gunpowder, the magnetic compass, and other discoveries—was not the birthplace of science. Why? Perhaps because the development of science in a culture is contingent on its world view. Europeans were convinced that the universe was the work of a single creator and that it operated in accordance with clearly defined divine laws; it was up to humans to puzzle them out. For the Chinese, however, the concepts of God and divine laws did not apply because, in their view, everything in the universe stemmed from the reciprocal action of yin and yang. Theoretical science was superfluous. The Chinese philosopher pictured in this 19th-century embroidery (left) is contemplating the yin-yang symbol.

The "Greek Miracle" Got Under Way in Ionia, on the Coast of Asia Minor, in the 6th Century BC

For the ancient Greeks, it was not enough to observe natural phenomena uncomprehendingly and blindly defer to the gods. The Greeks felt that the constituent parts of the universe were governed by laws that could be grasped by the human mind. Mortals could share in divine knowledge. The scientific method as we practice it today took shape in ancient Greece over the course of eight centuries. The notion that the only way to divine the harmony of the cosmos was to observe and measure celestial motion—the motion of the planets in particular—gained currency during this time.

The Greek World View: The Earth Lies at the Center of the Universe

It is easy to understand why the Greeks found geocentric cosmology so appealing. Night after night, we see celestial objects moving across the sky from east to west. Wouldn't it be natural to assume that the earth lies motionless at the center of the universe and that the sun, moon, planets, and stars revolve around it? In the 4th century BC, Plato (c. 428–348 BC) postulated a world system with the stationary earth as its hub and a huge outer sphere that carried the planets and stars in its daily revolution. But his two-sphere configuration could not account for a peculiar aberration in the motion of certain planets. As a rule, the planets and stars traveled together across the sky every night from east to west. But now and then, inexplicably, some planets seemed to drift

Before Plato's geocentric system—the one that culminated in the Ptolemaic model six centuries later (illustrated below left)—early Greek cosmology still had mythic overtones. Like the Babylonians, Greek astronomer Thales of Miletus (c. 625–c. 547 BC) believed that water was the primary substance of the universe. The earth, he claimed, was a flat disk floating on a primordial ocean and surmounted by a sky of water. Echoing the Chinese principles of yin and yang, astronomer Anaximander (610–c. 547 BC), argued that all things resulted from the interaction of opposites (hot and cold, light and dark). According to Greek philosopher Pythagoras (c. 580–c. 500 BC), numbers were the basis of all phenomena, and mathematical laws governed the universe.

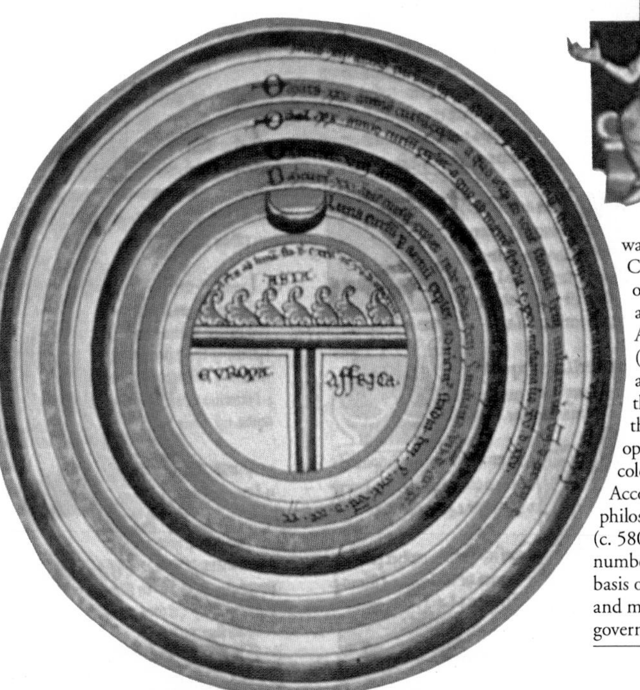

backward in relation to background stars. Today we know that retrograde motion is an illusion caused by the fact that we observe planetary motion from a planet that is itself in motion.

It never entered the mind of Eudoxus of Cnidus (c. 400–c. 350 BC), a young contemporary of Plato's, that the earth could actually move. But he did try to accommodate planetary retrogression to the motionless earth by developing from the two-sphere Platonic system a model with twenty-seven spheres. To the terrestrial and starry spheres he added planetary spheres, one for every known planet; to each of these were attached still other spheres whose rotation, combined with the rotation of the planetary spheres, explained why certain heavenly bodies "reversed" direction.

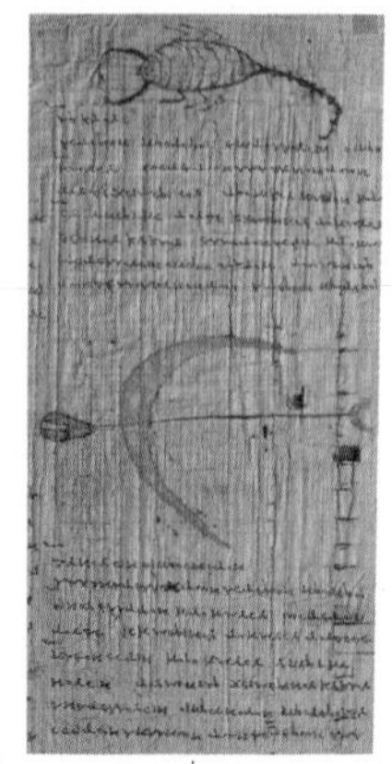

For Plato, the true world consisted of abstract geometric forms, or "ideas"; physical objects were mere shadows of reality. The key to understanding the universe lay in the mind, not the senses. Eudoxus took a different philosophical approach (drawings of the moon, sun, and signs of the zodiac, above and left, from a manuscript copy of his writings). He maintained that reality could not be perceived by reason alone; observation was necessary to show it the way.

"And God Created the Heaven and the Earth"

Eudoxus' multispherical model, later refined by Ptolemy (2nd century AD), went unchallenged until the 16th century; it did, after all, account for observed celestial motion. However, it lacked spiritual resonance. That was to be the contribution of Aristotle (384–22 BC), and of St. Thomas Aquinas (1225–74), sixteen hundred years after him.

About 350 BC, Aristotle separated the cosmos into two realms with the lunar sphere as the boundary between them. The earth and the moon lay in the changing, imperfect world of birth, decay, and death, that realm of earth, air, fire, and water where all objects naturally moved up and down. But in the "superlunary" stratum of the other planets, the sun, and the stars, all was unchanging and eternal. Everything there naturally moved in perfect circles, which was why the planetary spheres rotated perpetually around the earth.

The role God played in Aristotle's scheme was not

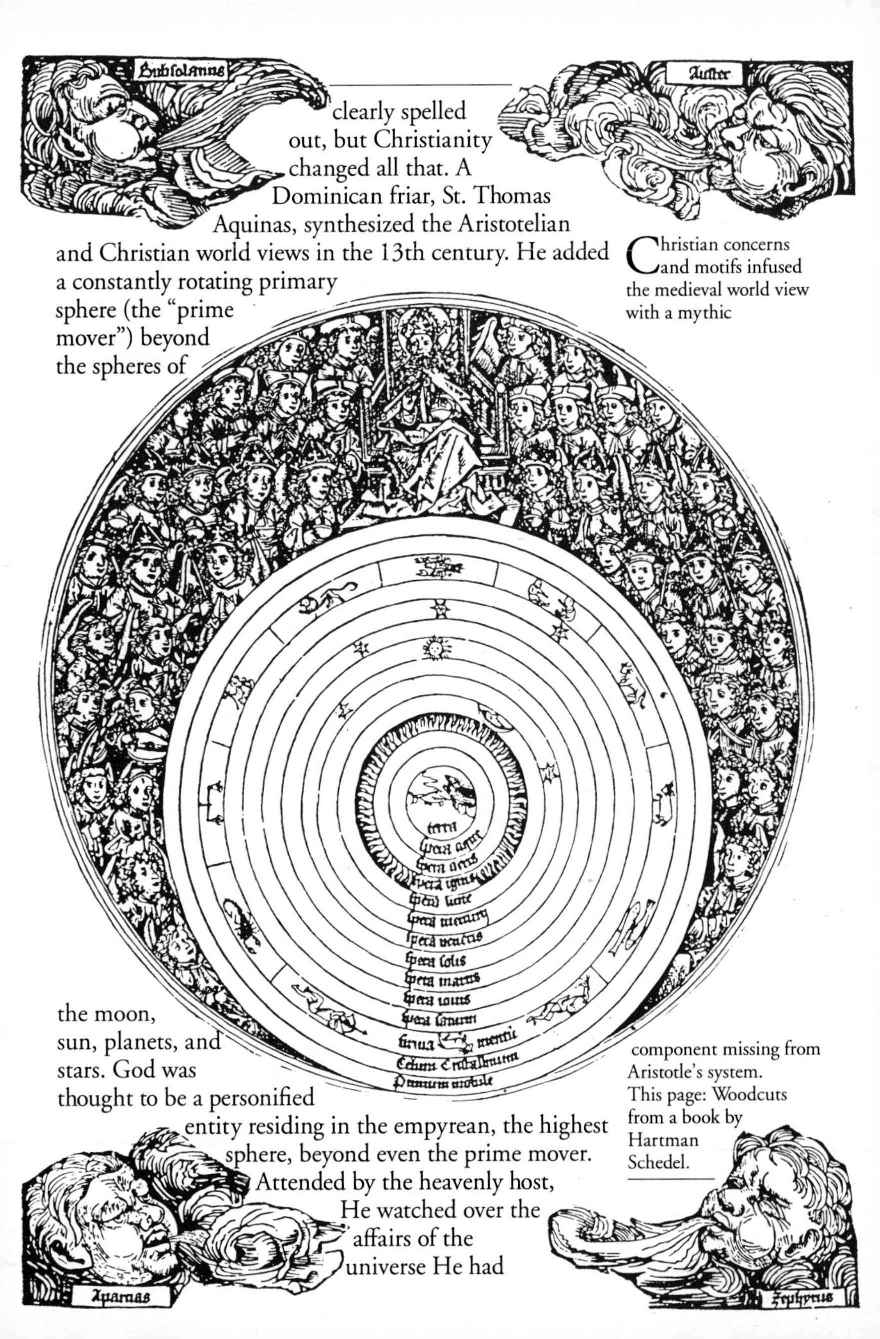

clearly spelled out, but Christianity changed all that. A Dominican friar, St. Thomas Aquinas, synthesized the Aristotelian and Christian world views in the 13th century. He added a constantly rotating primary sphere (the "prime mover") beyond the spheres of the moon, sun, planets, and stars. God was thought to be a personified entity residing in the empyrean, the highest sphere, beyond even the prime mover. Attended by the heavenly host, He watched over the affairs of the universe He had

Christian concerns and motifs infused the medieval world view with a mythic component missing from Aristotle's system. This page: Woodcuts from a book by Hartman Schedel.

created. The angels dwelled in the solar and planetary spheres, operating the machinery that moved the spheres in their orbits. Their degree of divinity diminished with the distance of their assigned dwelling place from the realm of God. Down in the sublunary stratum lay purgatory; earth, realm of humankind and mortality; and, deep below the surface, hell, dwelling place of devils and the souls of evildoers after their earthly existence.

Heliocentrism, or the Copernican Revolution

The universe revolved around the motionless earth for nearly two thousand years. Then, in 1543, a Polish astronomer named Nicolaus

Copernicus (1473–1543) published *De Revolutionibus Orbium Coelestium* (*On the Revolutions of the Celestial Spheres*), sparking an intellectual revolution whose repercussions are still being felt to this day. He removed the earth from its place of honor and enthroned the sun

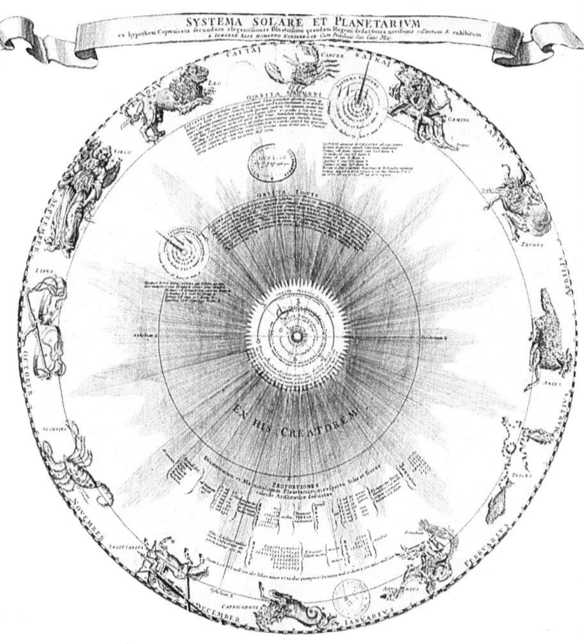

in its stead. He set the earth in motion on an annual circuit around the sun. It became just another planet.

The heliocentric theory dealt human pride a staggering blow. Humans lost their hegemony over the cosmos. The Lord had not singled them out for special treatment after all; the universe had not been created for them. The earth was now relegated to the superlunary realm of planetary spheres. But according to Aristotle, it was imperfect and changing. Did this mean that the superlunary spheres and the heavenly spheres beyond were likewise subject to change and decay? Could Aristotle have been wrong?

The universe had become a considerably vaster place, too, making the earth that much smaller and less

Scientists and nonspecialists alike were slow to be won over by the heliocentric system proposed by Nicolaus Copernicus. As the allegorical frontispiece (opposite) of Giovanni Battista Riccioli's *Almagestum Novum* attests, Aristotle's geocentric scheme (at the foot of the figure on the right) was no longer in the running by 1651. Yet, Tycho Brahe's compromise between the Aristotelian and Copernican models still outweighed heliocentrism.

By the early 18th century, Copernicus had triumphed; the sun now stood at the center of the zodiac (left, a 1700 print). Why didn't Copernicus incur the wrath of the church, which promoted the geocentric view? A clergyman himself, he prefaced his book with a statement that his system was postulated as a mathematical model, not scientific fact. Apparently this disclaimer satisfied ecclesiastical authorities.

significant. Before Copernicus, the entire universe was thought to extend no farther than the solar system; the outermost starry sphere lay just beyond the sphere of Saturn. Copernicus' universe was still finite and bounded by a starry sphere, but that outlying sphere became fixed. If night after night the stars appeared to move, it was not because the heavens rotate around the earth, but because the earth rotates on its axis once a day.

Postulating that the earth moves and the stars do not left Copernicus with no

alternative but to push the starry sphere very far out indeed, because, despite the earth's revolution around the sun, he noticed that the stars remained stubbornly fixed relative to one another. A nearby star observed at two different times during the earth's annual journey should appear to change position with respect to more distant stars. Yet, no such changes could be discerned. The stars, Copernicus concluded, must be much farther away from the earth than anyone had suspected.

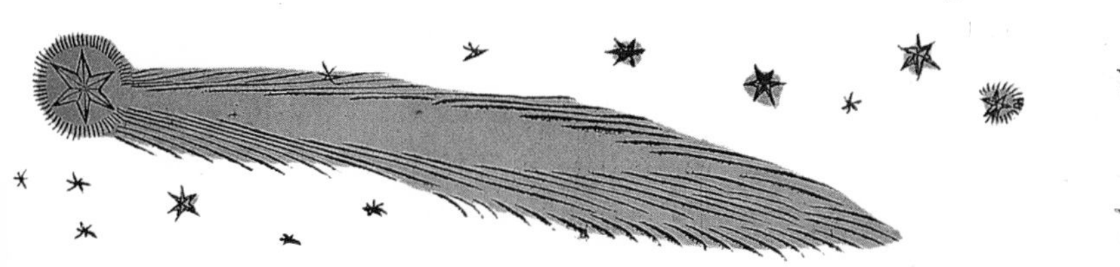

What Keeps Planets from Falling?

The Danish astronomer Tycho Brahe (1546–1601) heralded the next phase of the Copernican revolution by observing astronomical phenomena with a degree of accuracy never before achieved. In 1572 he spotted a new star in the constellation Cassiopeia, a star so bright it was visible during the day for a month. Unlike the planets, it did not change position in relation to distant stars; therefore, he reasoned, it must be very far away, well beyond the planetary spheres. A change had occurred in the heavens, further undermining confidence in the Aristotelian concept that the stars were perfect and unchanging. The "new star," we now know, was in fact a supernova, a dazzling explosion signaling, in a final burst of energy equal to that of billions of suns, the death of a massive star in our Milky Way.

Belief in Aristotle's perfect celestial spheres was shaken anew by the appearance of the Great Comet of 1577. Until then comets were thought to occur in the earth's atmosphere, like rainbows. Tycho Brahe proved that this could not be so. Since the comet changed position relative to distant stars, it was much closer to earth than the supernova was. But its motion across the sky was not as great as the moon's, so it had to lie beyond the lunar

King Frederick II of Denmark granted Tycho Brahe an entire island near Copenhagen for his astronomical studies. It was on Hven that Tycho built his main observatory, Uraniborg, and a smaller counterpart, Stjerneborg, for his assistants, here seen making preparations for the night in its dome-covered observation room (opposite below). One of them is holding a quadrant, for measuring altitudes; the telescope had not yet been invented. Under Tycho's supervision, Uraniborg became the leading observatory in Europe; from there he studied the Great Comet of 1577 (drawing, above). The planetary system he proposed was a compromise between Copernicus' heliocentric model and Aristotle's geocentric model: The planets are in orbit around the sun, but the sun, like the moon, is orbiting a motionless earth (opposite above).

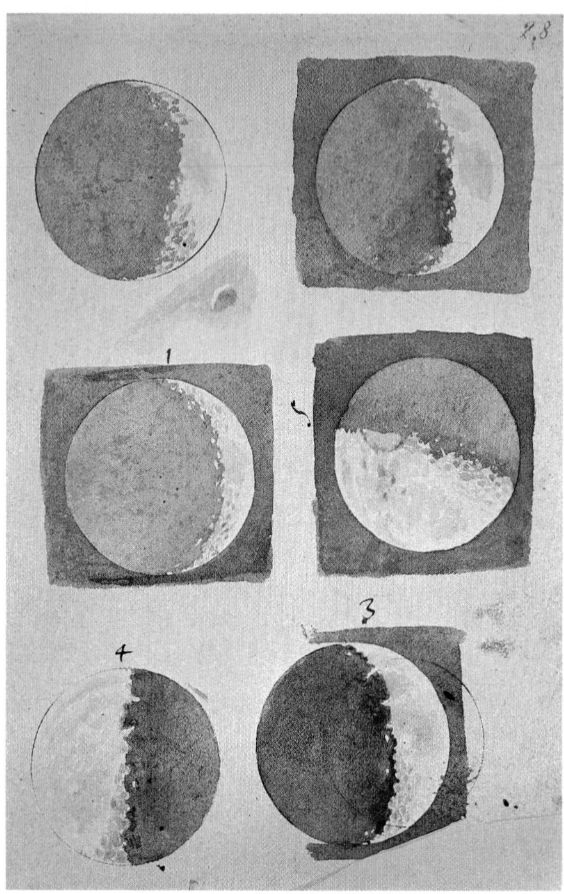

Father of experimental physics, Galileo began investigating the motion of falling bodies early in his career. He proved that every falling body, regardless of its weight, accelerates at exactly the same rate. Were it not for air drag, a feather and a lead artillery shell released simultaneously from the top of a tower would hit the ground at the same time. In 1609 he studied the phases of the moon and recorded his accurate observations in a series of detailed sketches (left and below).

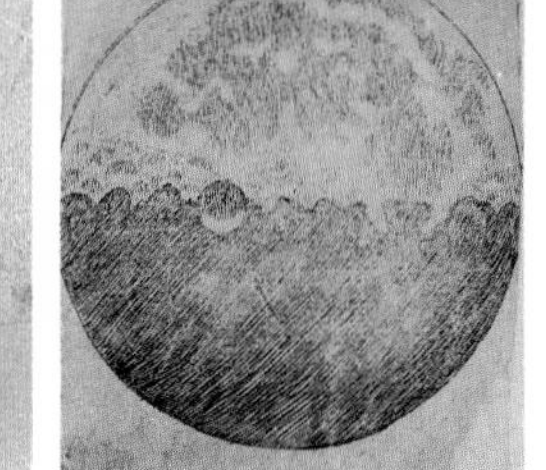

Galileo discovered that Venus exhibits phases like the moon's—a result of reflected sunlight—and concluded that the planet must be in orbit around the sun.

sphere. Its location, he concluded, was somewhere in the region of the planetary spheres.

Tycho calculated that the comet's orbit was oval, not circular—a direct challenge to the notion that all celestial bodies naturally moved in perfect circles. Furthermore, an oval orbit would send the comet smashing through the crystalline planetary spheres. That defied logic if one assumed that such spheres really existed, and so Tycho was forced to conclude that they were mere figments of the imagination. But if the planets were not attached to solid spheres, what kept them from falling? What held them up in the sky?

Galileo Reconciled Heaven and Earth

In the Aristotelian system, terrestrial and celestial phenomena are governed by different natural laws: Bodies move along straight lines on the earth but travel in circles in the heavens. That long-standing belief was toppled by the Italian mathematician, astronomer, and physicist Galileo Galilei (1564–1642). He maintained that the realms of the earth and heavens are, in fact, one. In other words, the same natural laws—laws the human mind could puzzle out through painstaking observation—must govern everything in the universe. In 1609–10 Galileo became the first person ever to turn a

telescope skyward. It revealed fresh evidence of celestial "imperfection": Mountains broke the surface of the moon, dark spots blemished the surface of the sun. His discovery of four satellites orbiting Jupiter further discredited the proposition that everything revolves around the earth.

In 1632 Galileo openly declared his support for heliocentrism in the landmark *Dialogo Dei Due Massimi Systemi del Mondo* (*Dialogue Concerning the Two Chief World Systems*). It was more than church authorities

Suspecting he had made a momentous discovery, Galileo (above) immediately sent a coded message to Kepler in Prague: "Cynthiae figuras aemulatur mater amorum" ("Shapes of the moon are imitated by the mother of loves").

TABVLA III. ORBIVM PLANETARVM DIMENSIONES,
REGVLARIA CORPORA
ILLVSTRISS. PRINCIPI, AC DNO. DNO.
TENBERGICO, ET TECCIO, COMITI MONTIS

could bear. They placed him under house arrest until 1642; his book remained on the list of banned publications until 1835, and his conviction was not overturned until 1992. The divorce between science and religion was now complete.

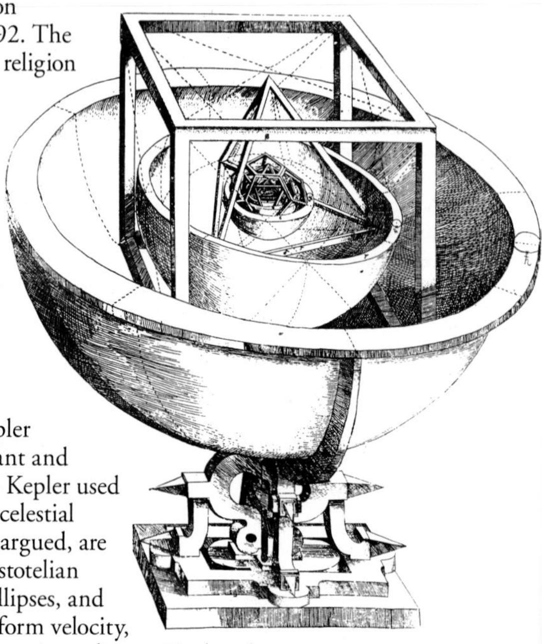

Scientific Explanations of How the Universe Works Still Rest on Laws Kepler and Newton Formulated in the 17th Century

Tycho Brahe's incomparably accurate positional measurements of planetary motion proved invaluable to German astronomer and mathematician Johannes Kepler (1571–1630), Tycho's assistant and successor in Prague. In 1606 Kepler used them to unlock the secret of celestial motion. Planetary orbits, he argued, are not perfect circles, as the Aristotelian world view demanded, but ellipses, and planets do not move at a uniform velocity, the perfect Aristotelian motion. According to Kepler, the closer a planet is to the sun, the faster it travels; the farther away, the more slowly it travels.

However, Kepler's mathematical laws of planetary motion left unresolved the dilemma Tycho faced when he did away with solid planetary spheres. What holds the planets in their orbits? Why don't they fall toward the sun? What propelled them around the sun now that there were no angels left to push them along?

Isaac Newton (1642–1727) answered these questions in 1666, consigning the Aristotelian heaven-earth dichotomy to the dustbin of history once and for all. As the English physicist and mathematician saw it, the motion of a ripe apple falling from a tree and the motion

Before he formulated the laws of planetary motion, Johannes Kepler assumed that geometry governed the universe and that the spheres of the six known planets were nested in the five perfect solids of Plato and Pythagoras, the outermost being the cube (illustrated in an engraving from his *Mysterium Cosmographicum,* above).

ET DISTANTIAS PER QVINQVE GEOMETRICA EXHIBENS.

FRIDERICO, DVCI WIR: BELGARVM, ETC. CONSECRATA..

of the moon around the earth are subject to one and the same force: universal gravitation. An apple flung into the air naturally describes a curve in space; by the same token, the moon needs nothing supernatural to drive it in its orbit.

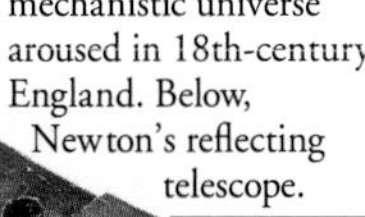

The Orrery by Joseph Wright of Derby (1766, bottom) exemplifies the keen interest Newton's mechanistic universe aroused in 18th-century England. Below, Newton's reflecting telescope.

The Unessential Hypothesis of God

According to Newton, a universe governed by universal gravitation must be infinite. If it were finite, the force of gravity would cause everything in it to collapse into some central point—an outcome not consistent with the observed universe. The Newtonian universe runs like clockwork. It is deterministic: Everything in it is strictly regulated by precise laws. There was no further need for God to intervene in human affairs. He had wound the cosmic "spring"; all He had to do was watch things unfold from afar.

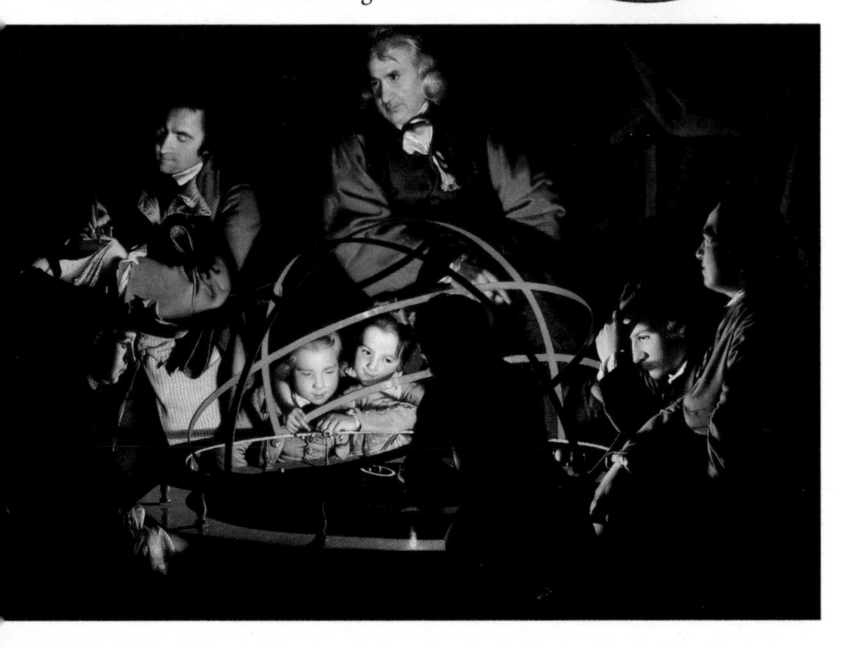

With a mirror 72 inches wide, the reflecting telescope Irish astronomer William Parsons, third earl of Rosse (1800–67), built for himself was for a time the world's largest. It is shown being constructed in this contemporary print. Some nebulas, he discovered, had a spiral structure. His sketches of Messier 51 (the Whirlpool Galaxy, above) and Messier 99 (opposite below) are splendid. The Messier catalogue was prepared in 1781.

In fact, God became so remote that by the 18th century French astronomer and mathematician Pierre-Simon de Laplace (1749–1827) felt he could dispense with Him altogether. After reading a presentation copy of Laplace's *Traité de Méchanique Céleste* (*Treatise on Celestial Mechanics*), Napoléon Bonaparte took him to task for failing to mention the Great Architect even once.

"I have no need of that hypothesis," Laplace retorted dryly.

Thus, by the 19th century, people in the Western world had been dwarfed to insignificance by infinity, distanced from God, and consigned to a mechanistic, deterministic universe. Still, they could cling to the comforting belief that their lineage reached back to Adam and Eve, whom God had specially created to "have dominion over every living thing that moveth upon the earth."

Charles Darwin (1809–82) shattered that last illusion when he published *On The Origin of Species by Means of Natural Selection* in 1859. Our beginnings, the English naturalist maintained, were far less noble than previously supposed: People were descended from the apes by way of reptiles, fishes, and primitive cells.

By Kepler's and Newton's reckoning, the universe was 6000 years old, but geological evidence was starting to show that biological evolution had taken much longer—billions of years, in fact. The universe had already expanded in space. Now it was expanding in time.

The brilliant mathematician and astronomer Pierre-Simon de Laplace made significant contributions to understanding planetary motion. In addition to proposing a theory of how the solar system was formed, he was the first to postulate the existence of black holes, which he called *astres occlus,* or "dark stars."

The advent of large-aperture telescopes in the early 20th century made it possible to explore the heavens systematically. It soon became clear that our sun is only one of a hundred billion stars in the Milky Way, itself but one of hundreds of billions of galaxies spread throughout the universe.

CHAPTER II
THE REALM OF THE GALAXIES

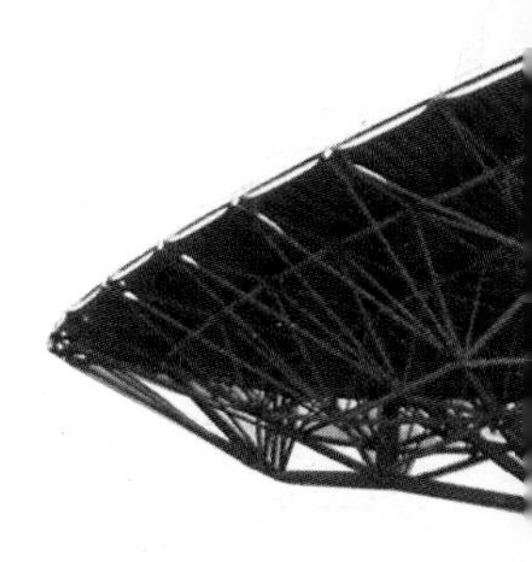

On clear winter nights, the Andromeda Galaxy, some 2.3 million light-years from earth, is visible to the unaided eye (left). Its central region is made up of old, yellow stars, while the spiral arms harbor young, blue stars. It is flanked by two dwarf satellite galaxies. The Very Large Array (VLA), near Socorro, New Mexico (partial view, right), is a set of twenty-seven movable, identical 85-foot dishes arranged on railroad tracks along the arms of a huge Y.

Light is the single most important communications link between human beings and the cosmos. It speeds information toward us at about 186,281 miles per second, the fastest velocity attainable in the universe.

But there is a limit to what the unaided eye can detect. Its light-gathering power is exceedingly low. Moreover, we cannot stare at the same image indefinitely: The human brain is organized in such a way that it must renew a visually transmitted image once every three-hundredths of a second. Consequently, the naked eye can make out only the brightest and closest celestial objects; the distant universe escapes us completely.

Modern technology has significantly changed the way astronomers work. Telescopes are now fully automated. Gone are the days when, as the illustration below attests, pointing telescopes and opening observation slits were manual procedures. Our romanticized image of scientists glued to eyepieces, struggling to stay awake as they sit shivering in the dark, is a holdover from another age.

Since Galileo, Telescopes Have Been Getting Bigger and Better

Telescopes have proved helpful in two ways. Their larger surface permits them to gather more light over a much larger area than the human eye can, and they can also "lock" onto an object for any desired length of time. This allows them to make out very faint celestial objects at great distances from earth, revealing the depths of the sky. Of course, they magnify images, too, thus bringing details into sharper focus.

The first generation of telescopes collected and refracted light through lenses, the way eyeglasses do. But the lens diameter of early refractors was limited to about 40 inches; anything bigger was simply too thick and heavy to be practical. Later astronomy was dominated by reflecting telescopes, which gather and focus light by means of a large parabola-shaped mirror. Early in the 20th century, two telescopes—the 60-inch and 100-inch mirrors built in 1908 and 1922,

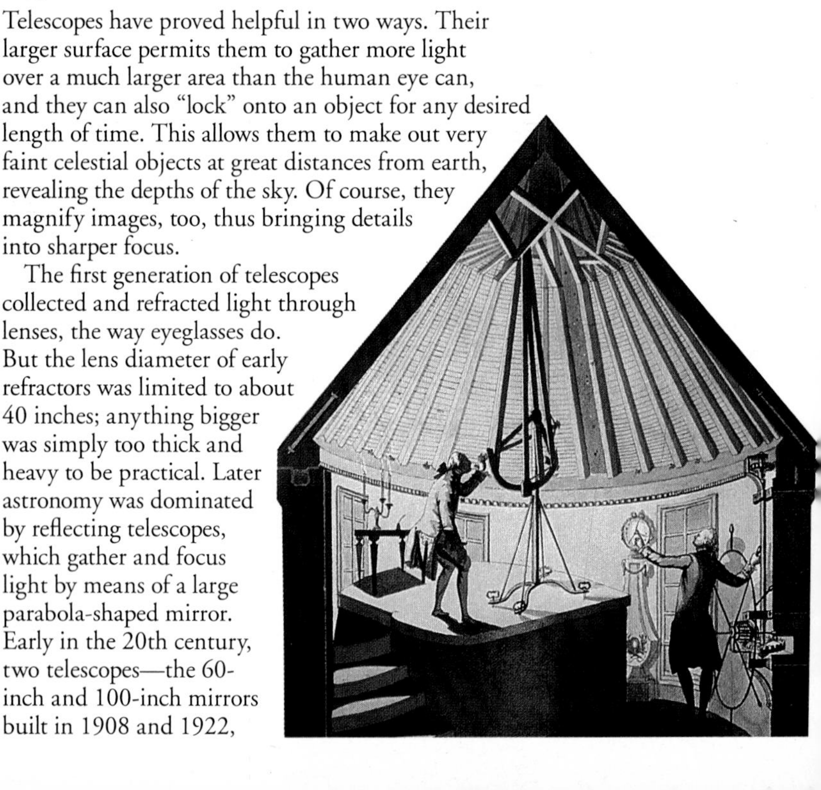

The gigantic 200-inch Hale reflector at Mount Palomar (left) was named after its builder, American astronomer George Ellery Hale. Like all modern telescopes, it is controlled by a powerful computer. After the telescope is pointed at a celestial object, the image, magnified thousands of times, shows up on a television monitor. The dome of the telescope (below) is as tall as a ten-story building.

respectively, at Mount Wilson in southern California—revolutionized the way we look at the universe. The 200-inch Hale reflector on Mount Palomar, California, opened in 1948 and was the world's biggest until Russia's 236-inch mirror at the Special Astrophysical Observatory in the Caucasus went into operation in 1976. The Hale telescope can detect objects 40 million times fainter than the dimmest star visible to the unaided eye.

On clear nights, some fifteen telescopes with diameters greater than 120 inches perched on remote mountaintops the world over—from Arizona to

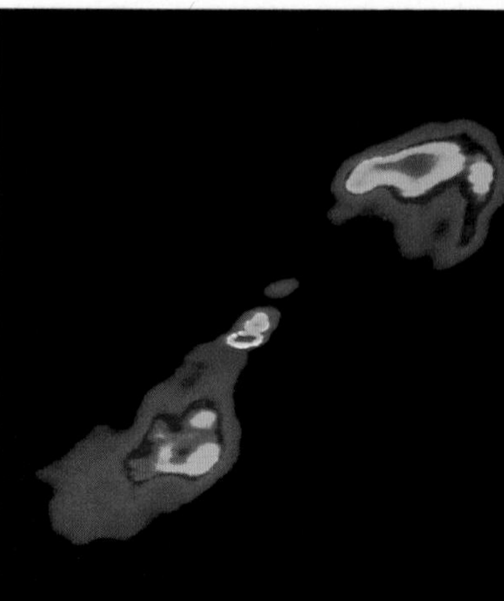

Hawaii, from the Caucasus to Chile—turn their mirrors skyward and capture the luminous messages the universe sends our way. Even bigger facilities are now being built or planned. Multimirror telescopes will have a total light-gathering power equivalent to that of single telescopes 400 to 600 inches in diameter and could attain a truly impressive range, ten times that of the Hale telescope.

Used in combination, the four 300-inch-diameter mirrors of European Southern Observatory's Very Large Telescope (VLT) now being built in northern Chile (below) have the power of a single 600-inch mirror.

The First Step Is to Gather Light, and the Next Is to Hold on to It Long Enough to Record Lasting Images

Early astronomers had to make do with sketching their observations. With the invention of photography in 1826, the images of thousands of stars could be permanently recorded on a single glass plate. Systematic sky surveys were now feasible. The process of building up images on large photographic plates over a period of hours substantially enhanced the power of telescopes to detect dimmer celestial objects; long exposures recorded extremely faint images.

This was the method of choice at most observatories until the 1970s, when it was superseded by ultra-sensitive electronic detectors (charge-coupled devices, or CCDs) able to gather as much light in half an hour

as a photographic plate could in an entire night.

Developed in the early 19th century by German physicist Joseph von Fraunhofer, spectroscopy—another breakthrough—made it possible to investigate the chemical composition and motion of stars and galaxies. Just as raindrops break up sunlight into the rainbow colors of the spectrum, spectroscopes break light from celestial objects into its various wavelengths.

Opposite and left: Images of Centaurus A, a galaxy 20 million light-years away in the direction of the constellation Centaurus, at visible, radio, and X-ray wavelengths (left to right). The huge dust lane cutting across the elliptical galaxy in the optical photograph looks dark because it absorbs visible light. The radio image reveals two long jets of radiation emanating from the center of the visible galaxy at right angles to the dust lane. At X-ray wavelengths, we see a single jet streaming downward from a bright core. Astronomers suspect that the source of these radio and X-ray jets is a supermassive black hole at the center of the visible galaxy.

"Invisible" Light

The telescopes described thus far collect visible light, the kind that can be perceived by the human eye. But it is important to remember that there exists a whole range of light we cannot see at all. Gamma rays and X rays, the most energetic (the highest frequency) forms of radiation, can pass right through human tissue. Though lower in frequency, ultraviolet radiation is sufficiently "hot" to

burn human skin and, in cases of overexposure to the sun, induce skin cancer. Next, in order of decreasing frequency, are the narrow band of visible light familiar to us all; infrared radiation; microwaves (the very same kind that cook our food); and, at the low-energy end of the spectrum, radio waves that carry programs from transmitting stations to our radio receivers and television sets.

From the standpoint of biological evolution, it was to our advantage to develop visual receptors sensitive to only the visible region of light, because the sun emits most of its energy at visible wavelengths.

The universe, however, uses the entire palette of light, seen and unseen, to paint the cosmic landscape. Confining ourselves to the range of the spectrum represented by visible light would be extremely limiting; it would be like looking at the universe with blinders on. Consider what would happen if suddenly our eyes were sensitive to just one color—say, for instance, red. We would see only scattered bits and pieces of the world around us.

Radio astronomy began in the 1950s following the development of radar during World War II. With the coming of the space age, telescopes could be borne aloft in balloons, rockets, and satellites. Astronomers finally got a peek at the universe above earth's atmosphere, which blocks gamma-ray, X-ray, ultraviolet, and infrared wavelengths.

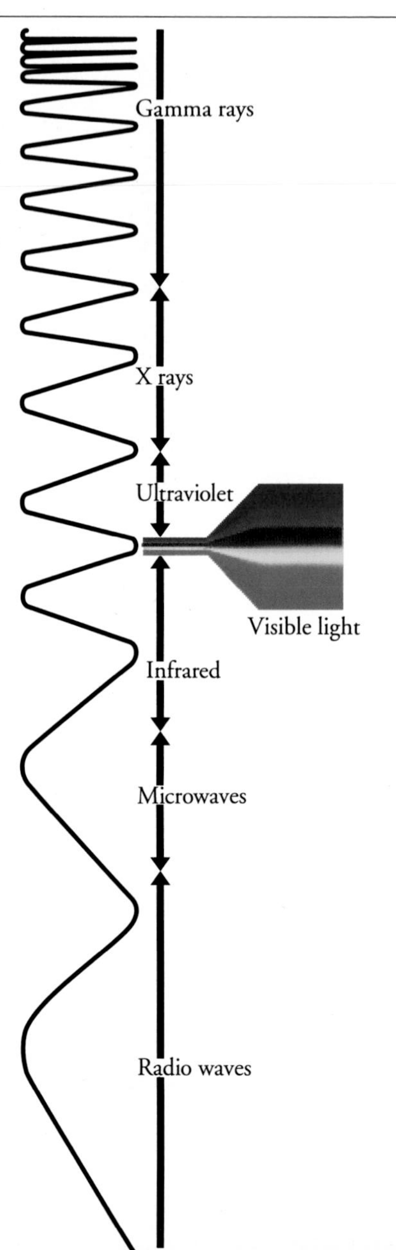

The Milky Way, a Galaxy 90,000 Light-Years Across

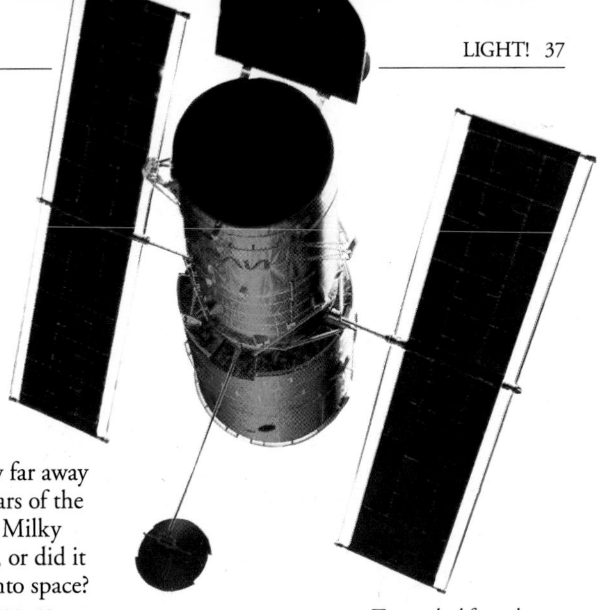

As Kepler's and Newton's successors set out to investigate the heavens with their telescopes, photographic plates, and spectroscopes, there were still many things about the universe that mystified them. How far away were the countless stars of the Milky Way? Did the Milky Way end somewhere, or did it extend indefinitely into space? If it was infinitely large, as Newton supposed, were the stars evenly distributed through its endless reaches? The answers were not obvious. The universe appeared to be spread out in two dimensions across the dome of the sky, as if someone with no knowledge of the elements of perspective had painted a landscape on a huge piece of canvas. The scientists' overriding concern was to unlock the secret of cosmic depth; things had to be put in their proper perspective.

Astronomers eagerly set about measuring distances from the earth to the stars. Even early in-depth surveys

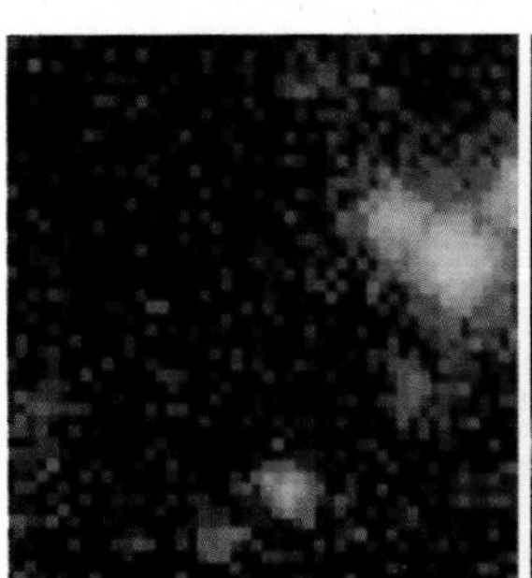

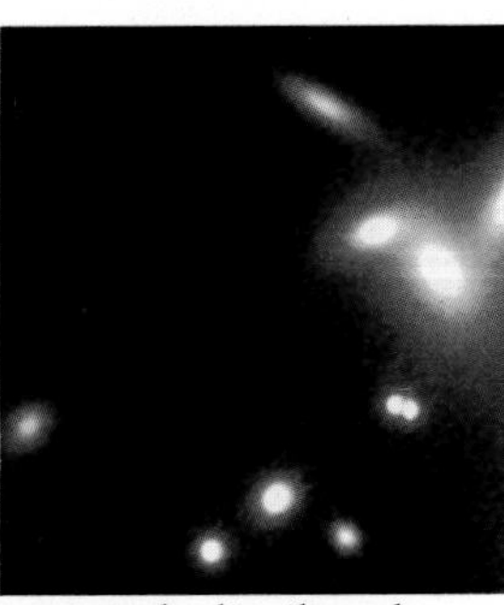

Launched from the space shuttle *Discovery* in April 1990, the 94-inch Hubble Space Telescope (above) was to have detected objects fifty times fainter than those visible by the most powerful ground-based telescopes and was to have improved resolution tenfold. Mount Palomar's 200-inch reflector images a distant galaxy cluster as a fuzzy blob of light (simulation, far left); HST was to have brought the same cluster into much sharper focus (left). Unfortunately, an optical flaw was found in Hubble's primary mirror after it was launched. NASA hopes to deploy shuttle astronauts to correct its myopia at the end of 1993.

of our little corner of the universe gave them some idea of how tiny our solar system really is and how much of space is a virtual vacuum. The sun, they learned, is 8 light-minutes from the earth; that is, its light takes 8 minutes to reach us. The size of our solar system can be reckoned in light-hours: Pluto, the most distant planet from the sun, is 5.2 light-hours from the earth. Stars are separated by light-years. The closest star after the sun is fully 4 light-years away. The heavens are vaster and emptier than anyone had imagined.

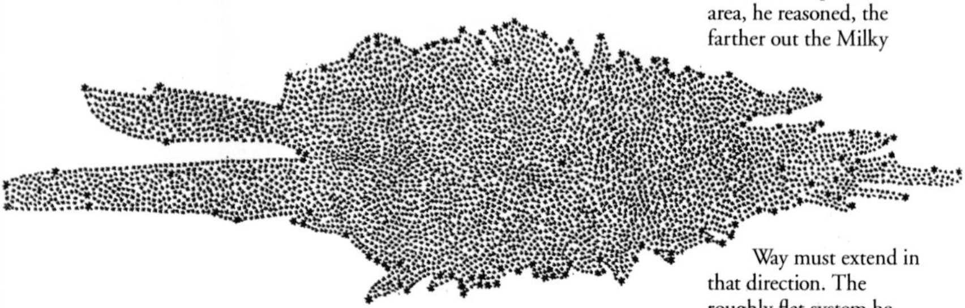

In the 1780s English astronomer William Herschel made a pioneering attempt to determine the shape of the Milky Way by counting stars visible in different areas of the sky. The more concentrated the stars in a particular area, he reasoned, the farther out the Milky Way must extend in that direction. The roughly flat system he came up with proved accurate; the jagged edge and centrally positioned sun did not. An illustration from his star catalogue (above).

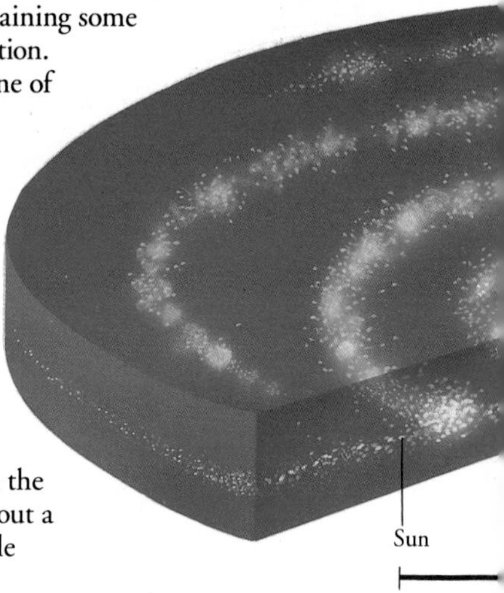

As space was probed ever more deeply, astronomers finally reached the edge of our galaxy, the Milky Way. It is not infinite in extent, as Newton had maintained, but a disk 90,000 light-years across containing some 100 billion stars held together by gravitation. Because our line of sight is along the plane of the galactic disk, we see the glow of the stars crowded in it as a spectacular band of pearly white arching across the night sky.

The solar system is one billionth the size of the galaxy. Measuring the breadth of the Milky Way from our speck of a planet was a stupendous feat roughly comparable to what an amoeba would achieve if it could somehow calculate the breadth of the Pacific Ocean. Moreover, the stars in the galactic disk, it was discovered, rotate about a galactic center. Contrary to what Aristotle believed, the stars do move after all.

Sun

The Solar System Was Banished to the Outskirts of the Galaxy

The sun was now known to be just one of 100 billion stars crowding the Milky Way. Humans consoled themselves in thinking that *their* star lay at the center of the galaxy. Then, American astronomer Harlow Shapley (1885–1972) disabused them of even that belief. Shapley investigated the distribution of globular star clusters—compact aggregations of up to 100,000 gravitationally bound stars—and found that they occupy a roughly spherical volume around the Milky Way. To his surprise, he discovered that the center of that sphere and the position of the sun did not coincide. The sun lay some 30,000 light-years away from the center, in the direction of the constellation Sagittarius. Shapley could only conclude that the sun is not located near the center of the Milky Way but on its outskirts, about a third of the way in from the edge.

The stars at the core of a globular cluster are so densely packed that an inhabitant would see ten thousand suns instead of one.

The bulge at the center of our galactic disk is a spherical system of a billion stars billions of years older than the sun.

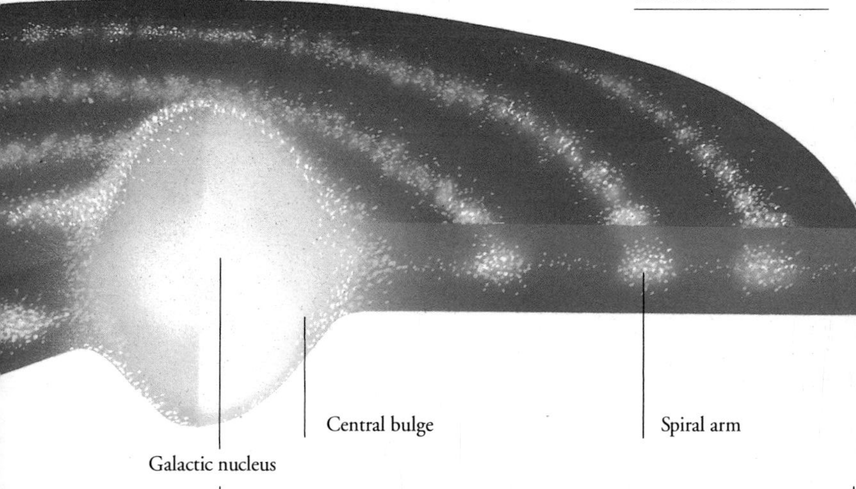

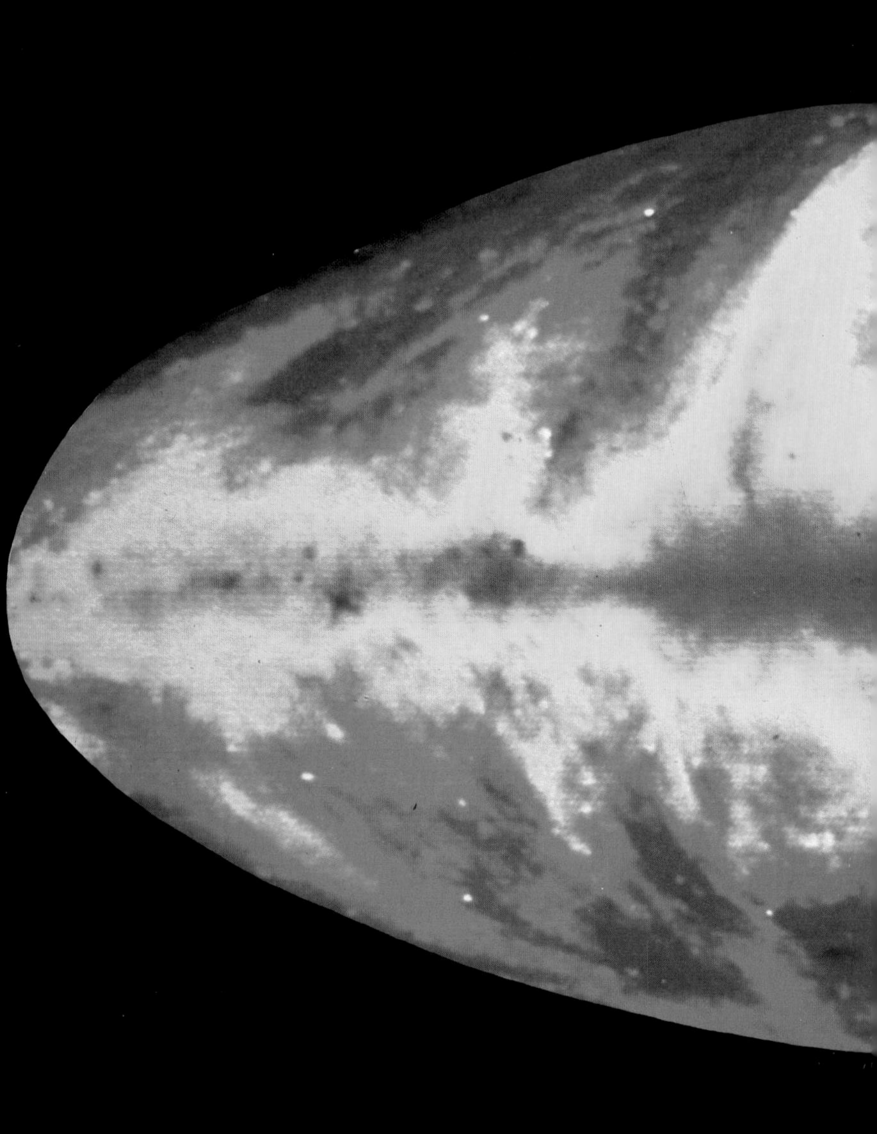

Portraits of the Milky Way: At Radio Wavelengths...

A radiophotograph of the Milky Way at a frequency of 408 MHz reveals not only the flattened galactic disk (orange) but also bands (yellow) arching gracefully from the disk toward the galactic halo.

Such a radio emission is thought to result when high-energy electrons, flung out into space at tremendous speed during the explosive death throes of massive stars (supernovas), interact with the Milky Way's magnetic field.

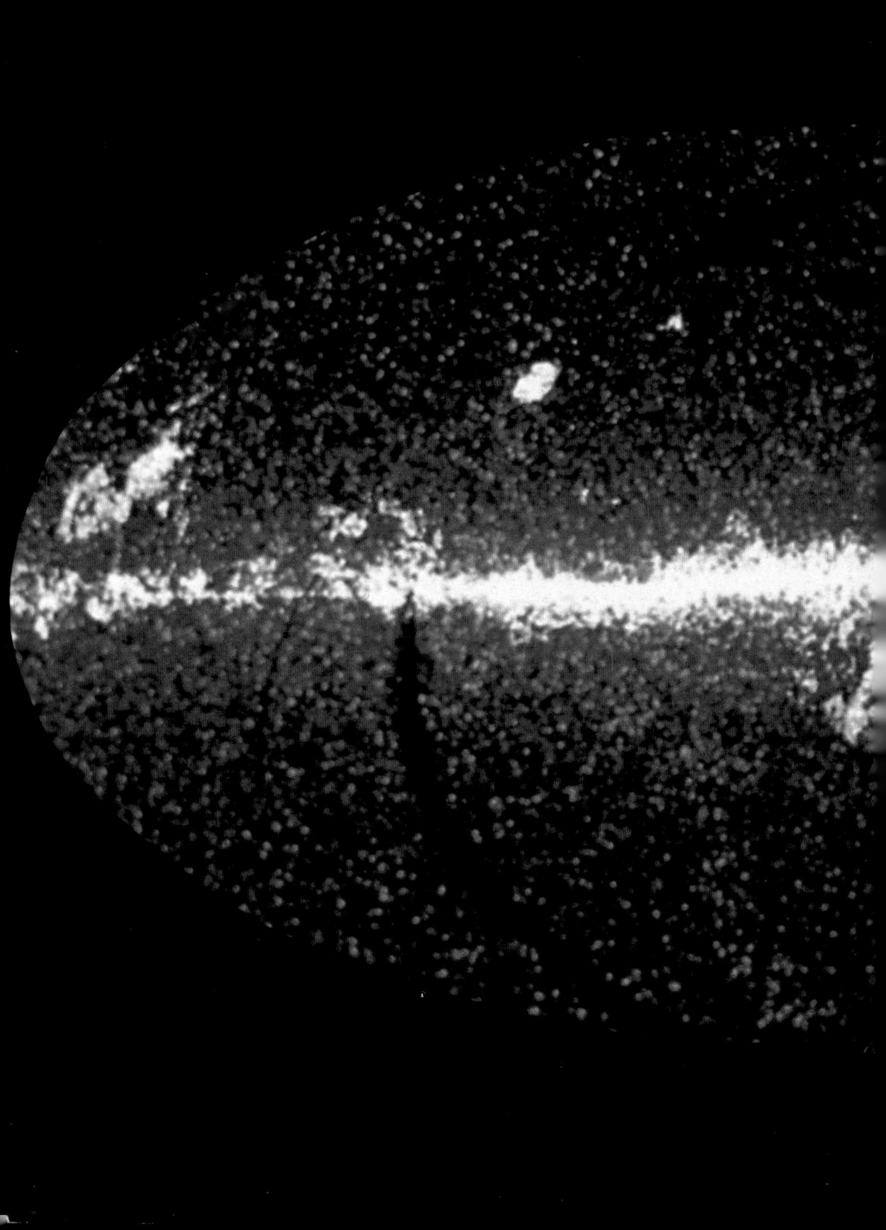

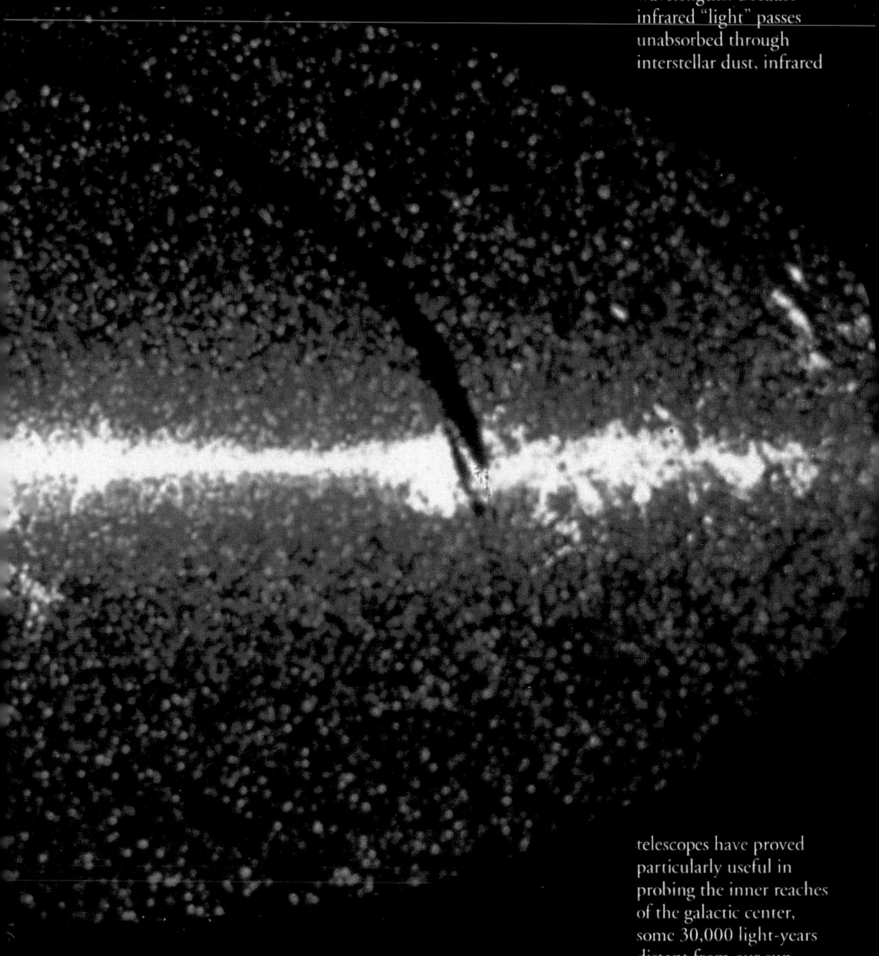

...in Infrared Light...

A detailed picture of the galactic disk emerges at infrared wavelengths. Because infrared "light" passes unabsorbed through interstellar dust, infrared telescopes have proved particularly useful in probing the inner reaches of the galactic center, some 30,000 light-years distant from our sun.

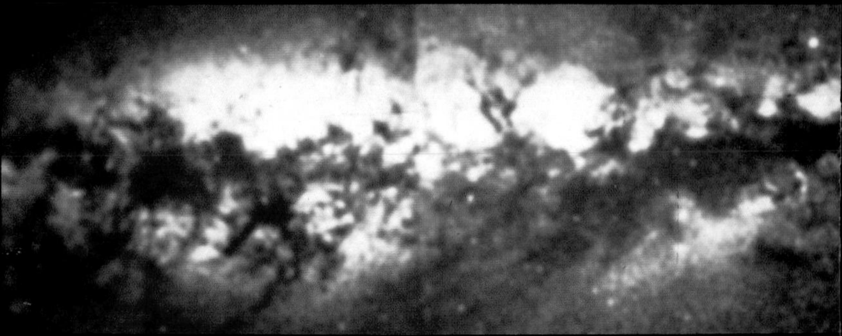

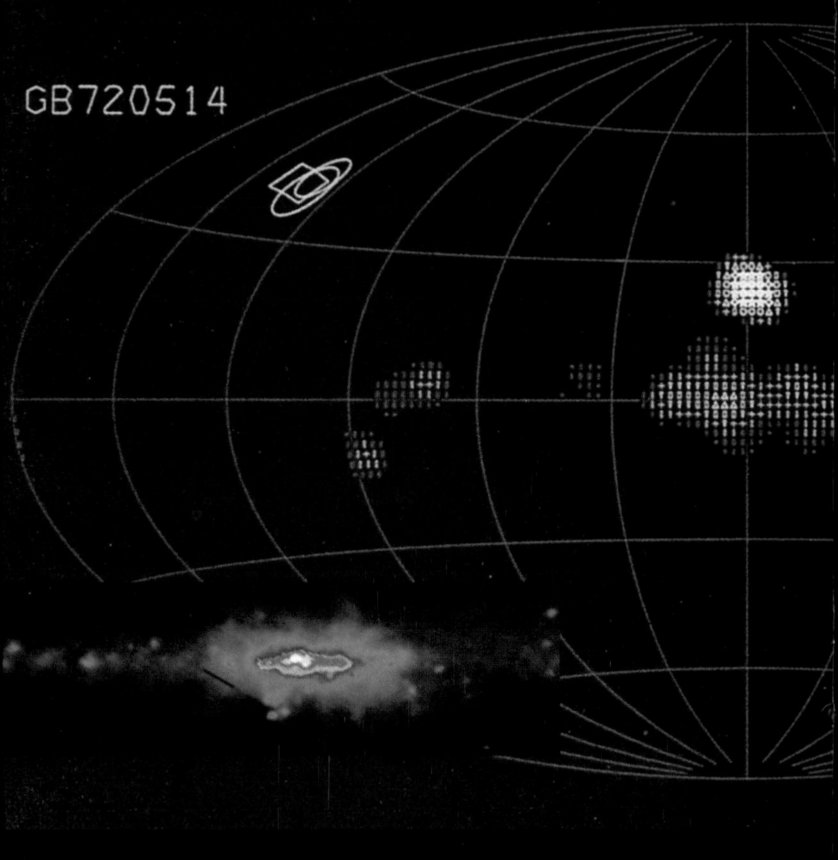

GB720514

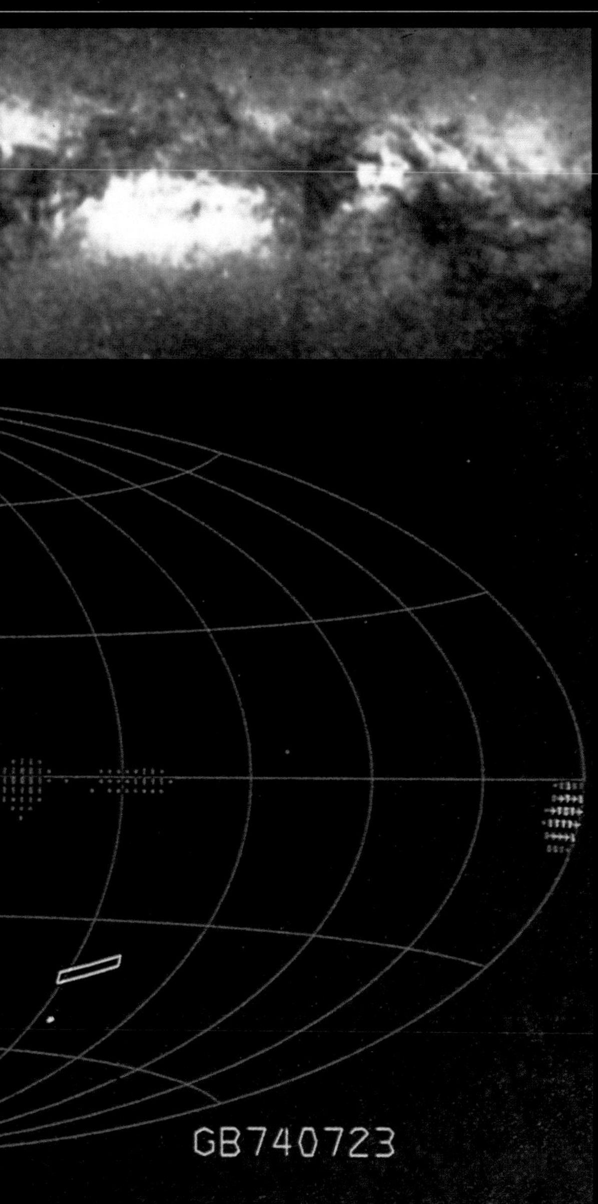

...and at Visible, X-ray, and Infrared Wavelengths

In visible light (top), parts of our galaxy's disk of stars are obscured by interstellar dust absorbing light from the stars behind them—the same dust that misled William Herschel as he tried to ascertain the shape of the Milky Way. Unaware that such interstellar material existed, he mistook what he saw in his telescope for the edges of the galaxy. The middle panel shows the Milky Way at X-ray wavelengths. Since high-energy radiation is emitted only when material is heated millions of degrees or more, the X-ray map pinpoints areas in the Milky Way where violent events have occurred or may now be unfolding. Observation of the galactic center (inset, opposite below) transmitted by the Infrared Astronomical Satellite (IRAS) shows a region where the stellar density is so high that the combined stellar light is two hundred times brighter than the full moon. Astronomers suspect that the nucleus of our galaxy harbors a black hole a million times more massive than the sun.

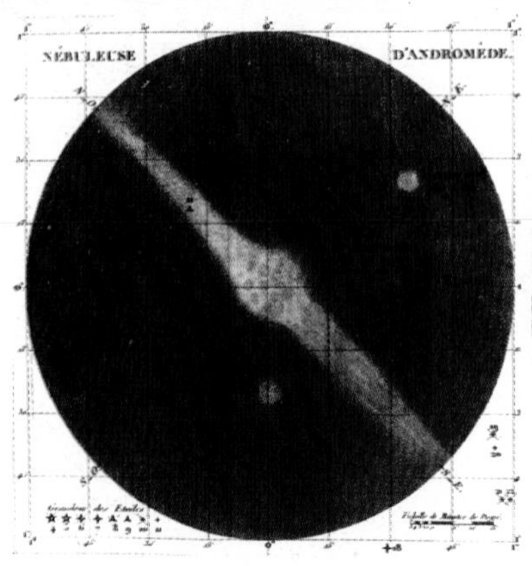

And Such a Little Galaxy at That!

That still left a fundamental issue unresolved. If the Milky Way ends somewhere, does its boundary mark the edge of the universe? Or does the universe extend even farther into space? Do other systems like our Milky Way lie beyond its confines? As early as 1775, German philosopher Immanuel Kant (1724–1804) theorized that star systems other than our own might exist. It seemed to some that the smudges of light (nebulas) English astronomer William Herschel (1738–1822) had recently discovered might be just such "island universes." Others were convinced that the universe and the Milky Way covered the same area and that nebulas must therefore be located within its bounds. The center of the universe, already shifted from the earth to the sun, shifted again, to the Milky Way Galaxy.

Astronomy's "great debate" raged. In 1924, American astronomer

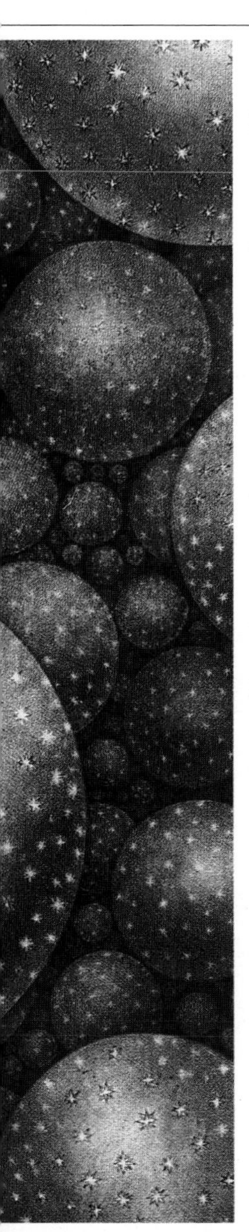

Edwin Hubble (1889–1953), who abandoned a law career to heed the call of the stars, used Mount Wilson Observatory's recently built 100-inch telescope to determine that the spiral nebula in the constellation Andromeda is far beyond the outermost reaches of our galaxy. It is now known to be 2.3 million light-years away. The nebula's light started out on its intergalactic voyage when humans first appeared on planet earth. The nebula in Andromeda, it turned out, closely resembles our own Milky Way Galaxy.

Kant's hunch about "island universes" was now scientific fact; suddenly there were more galaxies than anyone could count. The edge of the cosmos was pushed farther and farther out into space.

Our galaxy was becoming a speck in the vastness of the universe, just as our solar system had become a speck in the vastness of the Milky Way. Today we know that it is but one of the hundreds of billions of galaxies that exist—and it is an unremarkable one at that.

Charles Messier (1730–1817), who scanned the heavens hoping to spot new comets, sketched the Andromeda Nebula (opposite above) for his celebrated catalogue of nebulas (1781) but could only speculate about its properties. In *An Original Theory or New Hypothesis of the Universe* (1750), 18th-century English astronomer Thomas Wright surmised that nebulas were spherical counterparts of the Milky Way (engraving, left). In 1923 Edwin Hubble turned Wright's hunch into demonstrable fact when he identified a Cepheid variable in the Andromeda Nebula (above) and used this pulsating star as a celestial yardstick to calculate the nebula's distance from earth.

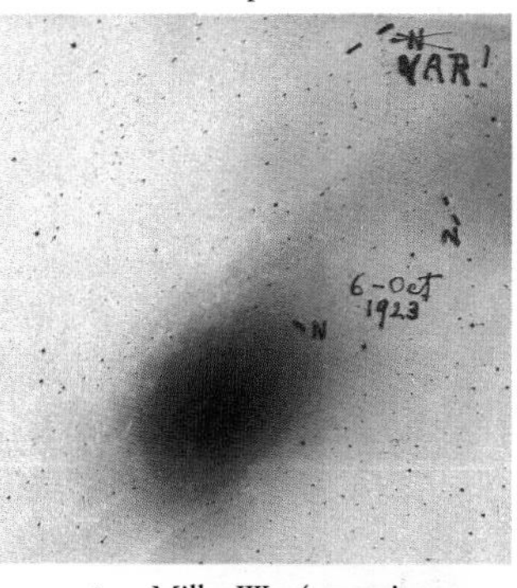

Classifying Galaxies: Elliptical, Spiral, and Irregular

Galaxies come in different shapes and sizes. Three out of ten are called elliptical galaxies because they are observed as oval patches of light. Six out of ten, including our Milky Way and its companion, the Andromeda Galaxy, resemble flattened disks with gracefully curved arms trailing outward. They are known as spiral galaxies. The remaining galaxies too indistinct in shape to be classified as spiral or elliptical are referred to as irregular galaxies.

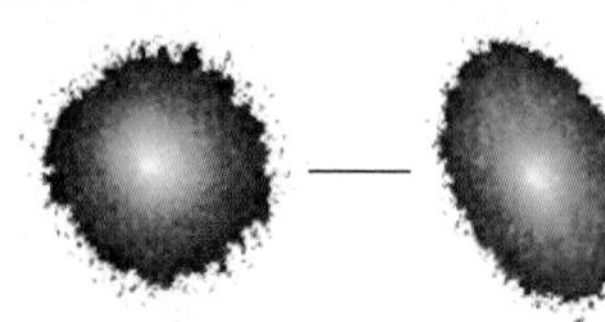

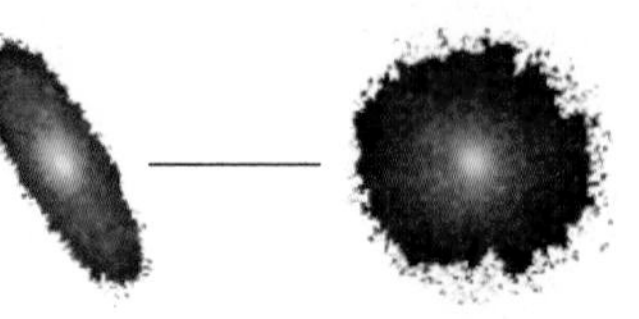

Hubble's "tuning fork" galaxy classification system (below), with elliptical galaxies along the handle and spiral galaxies along the prongs. The galaxy at their juncture is "lenticular," a transitional type combining observed characteristics of ellipticals (elliptical halo) and spirals (disk).

Why do these differences exist? The majority of galaxies are believed to have formed at about the same time, 2 to 3 billion years after the universe was born. Nascent galaxies consisted of clouds of hydrogen and helium, chemical elements forged during the first 3 minutes of the universe.

As these primordial clouds contracted under their own gravity, hundreds of billions of gas pockets of spherical shape began to condense out. The inward pressure of gravitation compressed them, and the temperature of their gaseous matter rose to tens of millions of degrees, triggering the fusion of hydrogen into helium and releasing tremendous amounts of

The Sombrero Galaxy in the Virgo cluster (above) is an excellent example of a lenticular galaxy.

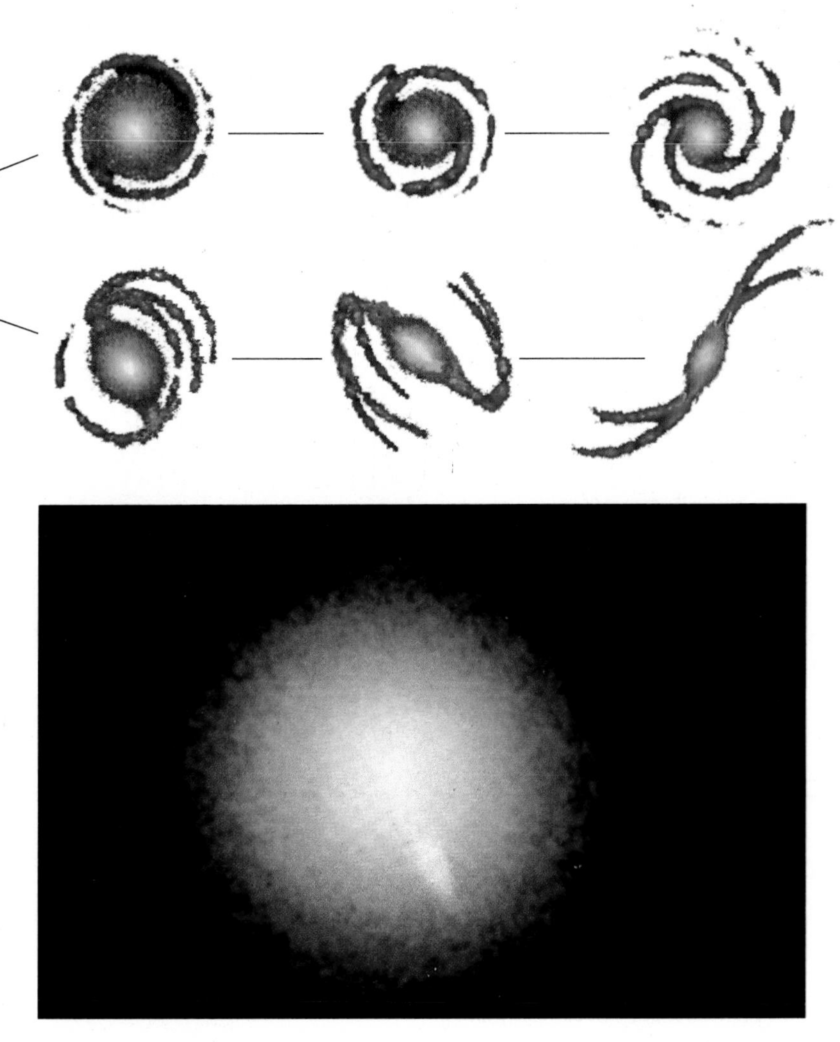

energy. The gas spheres thus "switched on" and became stars.

An embryonic galaxy's ultimate shape depended on how efficiently it could convert gaseous material into stars. A number of these protogalaxies took only a billion

Elliptical galaxies (M87, above) can be spherical or flattish. They are the predominant type in the central regions of galaxy clusters.

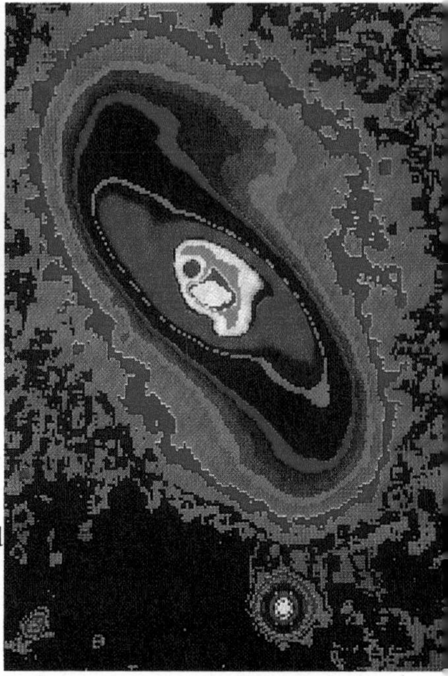

years to process all of the gaseous matter in them into stars. They evolved into ellipticals. Now lacking the raw material needed to breed new stars, they harbor old stars nearly as old as the universe itself.

Other protogalaxies were in less of a hurry and transformed no more than nine-tenths of their gas into stars. The remaining material was squashed into a flattened disk where gaseous matter continued to be converted into stars, but at a slower rate and mostly along the spiral arms. These star-forming regions, or stellar nurseries, are what give spiral galaxies their "youthful" appearance.

Still other protogalaxies processed their constituent gas at a positively sluggish pace. After 15 billion years of cosmic evolution, more than a fifth of their mass is still in a gaseous state. That is why irregular galaxies, though a thousand times less massive than their spiral counterparts, remain extremely prolific star breeders to this day.

Electronic devices convert optical images of celestial objects into sequences of numbers. Using powerful computers, astronomers can then process these digitalized images, eliminate clutter, and reconstruct natural-color images like that of spiral galaxy NGC 2997 (above left), with its bulge and well-delineated spiral arms.

Cosmic Collisions: Of "Accidents," Head-On Crashes, and Cannibalism

Galaxies possess a combination of inborn and acquired traits. Inborn properties, such as shape and mass, are determined at birth; other

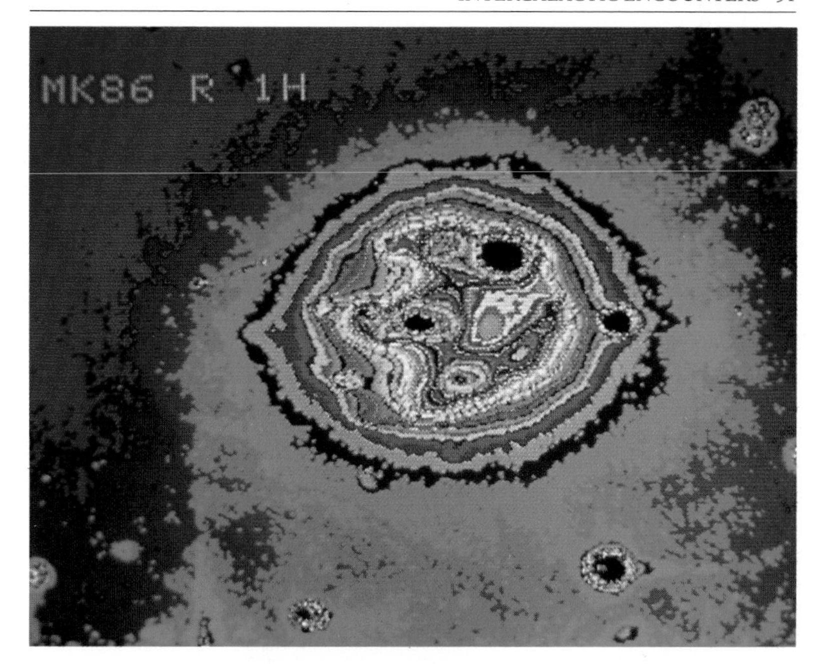

characteristics are acquired through interaction with their surroundings. That is because galaxies do not live in splendid isolation.

Gravity draws them together by the tens (groups) or thousands (clusters). Collisions between galaxies within the densely packed core of a cluster can radically alter their inborn properties. The more congested the galactic traffic, the higher the incidence of cosmic "accidents."

Most collisions do not involve direct impact. Intense gravitational forces can tear stars loose from the periphery of the colliding galaxies and hurl them into space, leaving the galaxies of a cluster awash in an intergalactic sea of stars. But that is usually the extent of the damage.

The aftermath of a head-on collision is more dramatic. If both of the galaxies happen to be spirals, the violent force of the impact can expel their gaseous disks into space. They merge, forming

False-color images of spiral galaxy NGC 89 (opposite right) and irregular galaxy MKN 86 (above). MKN 86, which stretches some 15,000 light-years across and is a billion times more massive than our sun, contains large amounts of hydrogen gas that is being actively converted into stars. The darkest areas (peak brightness) mark the location of huge stellar nurseries.

a brighter, more massive galaxy that changes into an elliptical for lack of gaseous material.

The Fate of the Milky Way

Such is the fate awaiting our own galaxy: Andromeda is due to collide with the Milky Way 3.7 billion years from now. The effects should not be dire, however, because the odds of our sun's crashing head-on into a member star of Andromeda are very low.

The realm of the galaxies is also rife with fierce cannibalism. Gravitational forces exerted by the biggest and most massive galaxies can affect the motion of less massive counterparts. Smaller neighbors that stray within range can find themselves drawn into a gradual spiral toward the "cannibals," which consume them and increase in size and mass the more galaxies they "ingest."

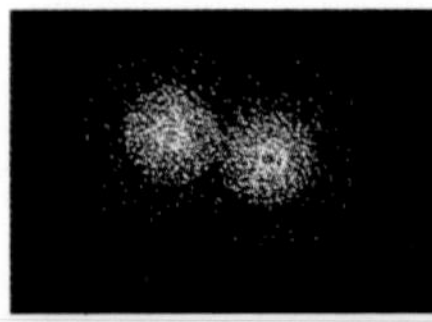

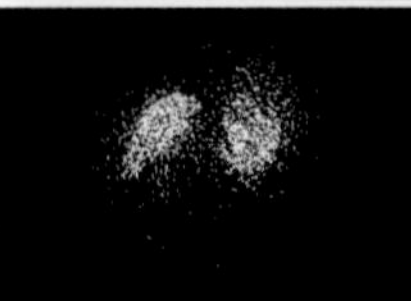

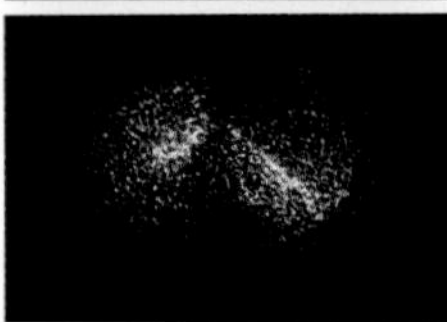

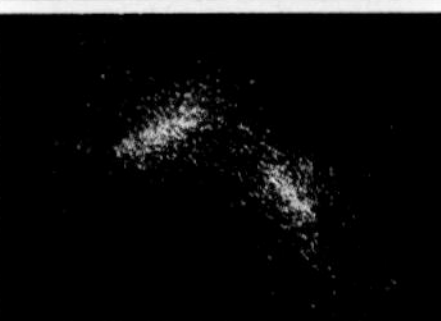

Astrophysicists use computer simulations to study galactic collisions and cannibalism. They can construct two galaxies, fling them at each other, and ask the computer to calculate the outcome, say, every 200 million years. This sequence of five images (left) graphically represents how two galaxies might interact over a period of a billion years. Because distances between individual stars average 3 light-years, the galaxies pass through each other. But gravitational forces can strip stars away from galaxies and create long streamers similar to the "tail" extending from interacting galaxies NGC 4676A and B, commonly known as the Mice Galaxy (below). The cores of the colliding galaxies are clearly visible in this image.

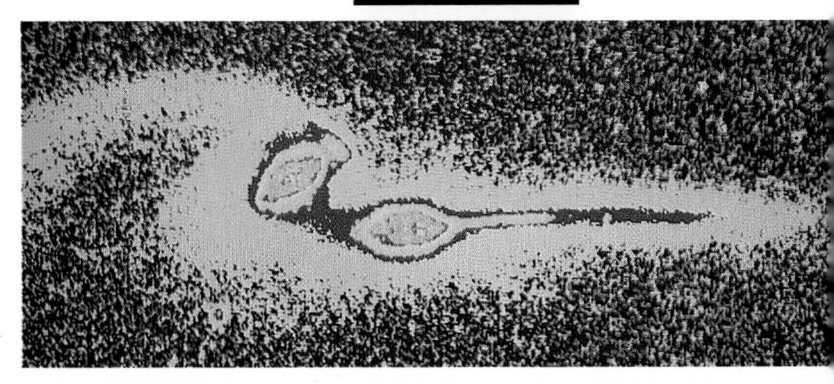

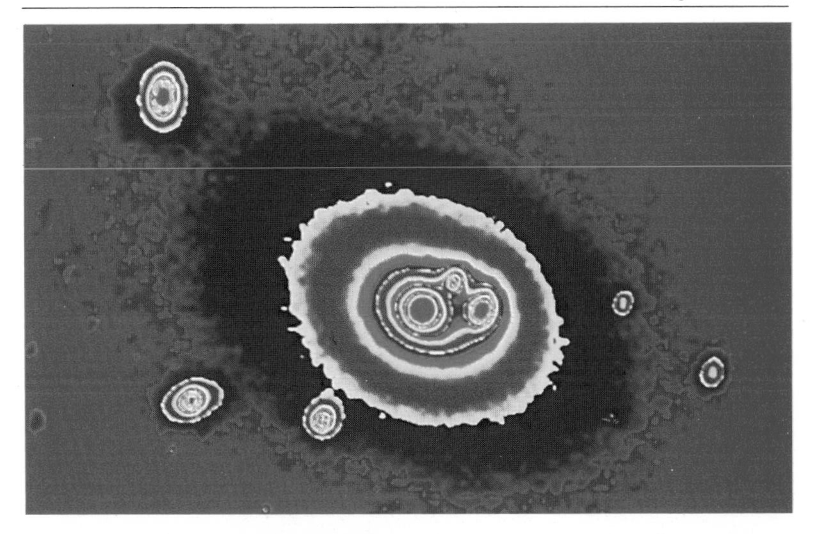

Computer processing enhances details deep inside the "belly" of a cannibal elliptical galaxy in the cluster Abell 2199 (above). The bright circular areas are galaxy remnants that have not yet been thoroughly "digested."

Quasars: Brightness Undreamed Of

The "eternal silence of infinite space" that filled French philosopher Blaise Pascal with such awe has been shattered by sound and fury. Modern telescopes sensitive to the entire range of incoming electromagnetic radiation tell of unimaginably violent events within the cores of certain galaxies. The most violent phenomena of them all may be quasars, which were initially identified as starlike objects in the Milky Way (from the term "quasi-star"). But when astronomers measured the distance to one of these intensely bright points of light they were stunned to learn that it lay some 13 billion light-years away, at the outermost reaches of the universe.

But how could an object so incredibly remote still radiate as much light as a star? There was only one possible explanation: Its intrinsic brightness must be enormous. Observations confirmed that quasars can emit as much energy as an entire galaxy and are as luminous as 100 billion suns combined. More astonishing yet, the underlying source of this stupendous outpouring is not much bigger than our solar system.

The Belly of a Black Hole

How can something so compact produce such unbelievable amounts of energy? No one knows for sure, but many astrophysicists suspect that quasars are born in galaxies that harbor monsters in their cores: supermassive, voracious black holes a billion times more massive than the sun, consuming all the nearby stars from the host galaxy. Black holes are regions of space with a gravitational pull so strong they trap even light, which travels at the fastest speed possible in the universe. Because they do not emit light, they cannot be seen and are therefore "black."

The gravitational forces of a black hole draw originally spherical stars into long filaments of material and funnel their shattered remains toward the yawning abyss at immense velocities. As the swirling particles of infalling gas get hotter, they pour out intense radiation before crossing the point of no return, the threshold beyond which the emission of radiation becomes undetectable.

You might say that quasar light is the swan song of shredded starstuff just before it winks out for all time.

The most distant and most luminous objects in the universe (left, an X-ray image of quasar 3C273), quasars pour out tremendous amounts of radiation at radio, infrared, visible, ultraviolet, X-ray, and gamma-ray wavelengths. The powerhouses generating their unbelievable energy are thought to be supermassive black holes consuming stars and gas from host galaxies. Quasars are so bright they outshine their host galaxies; that is why they are starlike in appearance.

In astronomy, looking out into space is the same as looking back in time. Windows on the universe when it was still very young, quasars tell us that galaxies started forming only about 2 billion years after the birth of the universe. Most quasars have since flickered out, probably for lack of stars, gas, and dust to feed the voracious black holes at their cores.

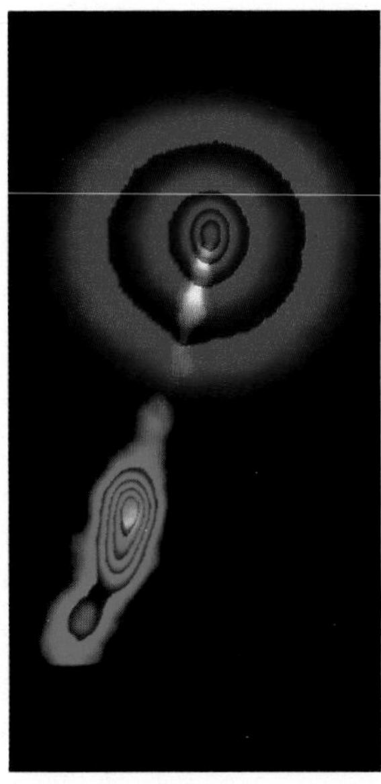

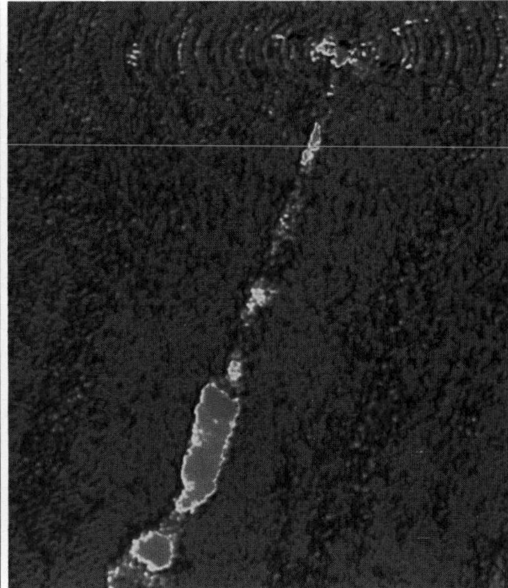

These radio (right) and computer-enhanced optical (left) images show a luminous jet extending 6500 light-years from the core of M87, a giant elliptical galaxy 50 million light-years away, near the center of the Virgo cluster. High-energy electrons caught in the intergalactic magnetic field create the radio emission from the jet. The suspected source of these fast-moving particles is a supermassive black hole—a billion times more massive than our sun—in the core of M87.

But quasar-core galaxies do not have a monopoly on these monsters.

Other types of galaxies with extremely luminous cores, so-called active galaxies, also keep the black holes lurking inside them supplied with stellar food. Because these black holes are ten to a hundred times less massive than the ones in quasars, they are less voracious. Still, they are strong enough to waylay the stars of a host galaxy, turning them into long, thin spaghetti-like filaments of matter that break apart and pour out intense radiation across the entire electromagnetic spectrum, from gamma rays and X rays all the way to microwaves and radio waves.

The Fabric of the Universe: Groups, Clusters, and Superclusters

Galaxies are grouped into communities. Our own galaxy is a member of what is called the Local Group, which includes the Milky Way, the Andromeda Galaxy, and about fifteen other dwarf galaxies, including

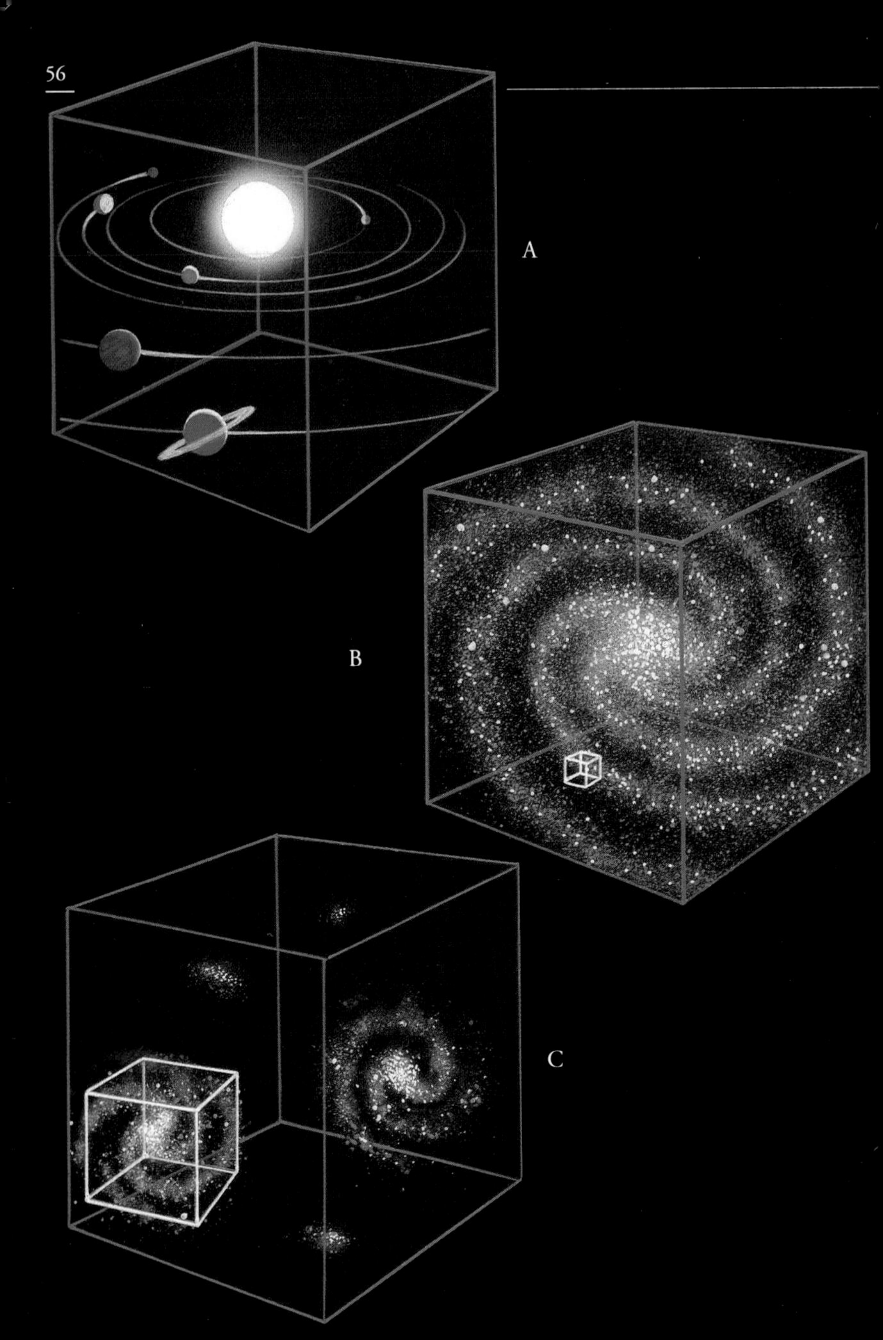
A
B
C

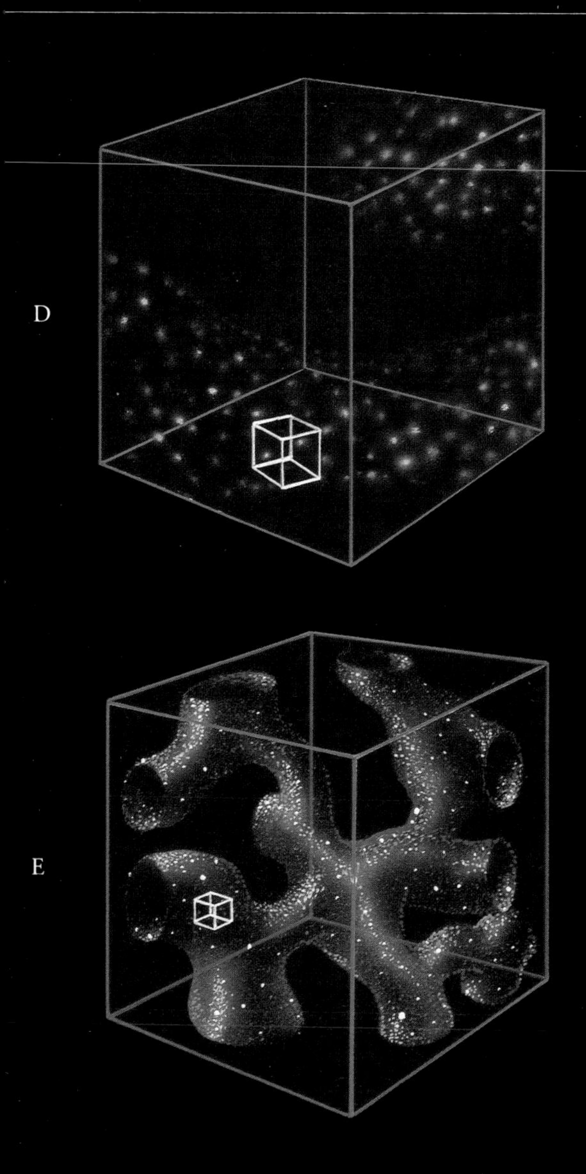

In-depth surveys of the universe show that the solar system to which our planet belongs plays but a tiny part in a stupendous cosmic ballet. Nestled in a solar system 10.4 light-hours in diameter, the earth whisks us through space at a speed of about 20 miles/second on its annual journey around the sun (A). The solar system orbits the galactic center at 145 miles/second (B). Our galaxy is edging toward its companion, Andromeda, at 55 miles/second; they are members of the Local Group, which is about 10 million light-years across (C). The Local Group, in turn, is streaking at 375 miles/second toward the Virgo cluster of galaxies in the Local Supercluster and the Hydra-Centaurus supercluster, which are spread some 60 million light-years across space (D). The ballet does not end there. The Virgo cluster and the Hydra-Centaurus supercluster are themselves being drawn toward another giant aggregation of galaxies astronomers refer to as the Great Attractor. These clusters and superclusters form unbelievably huge walls and filaments stretching hundreds of millions of light-years across space (E).

the Milky Way's dwarf satellites, the Large and Small Magellanic Clouds. Galaxy groups can spread up to 13 million light-years across space, that is, 130 times the diameter of a single galaxy. If galaxies are the houses of the universe, then groups are its villages.

The next larger units are clusters, the cities of the cosmos. These conglomerations, which number in the thousands, measure some 60 million light-years across and contain hundreds of trillions of stars.

The structure of the universe does not end there. Clusters are gathered into even larger superclusters, cosmic metropolises of millions of billions of stars sprawled out over hundreds of millions of light-years. The Local Group is part of a Local Supercluster containing ten or so individual groups and clusters.

In the photograph at left, all galactic types—including a spiral (below right) and two giant ellipticals, M84 (partly cut off, extreme left) and M86—are represented in the center of the Virgo cluster, a system of some thousand galaxies about 50 million light-years from the earth. It was in this cluster, in 1933, that Swiss astronomer Fritz Zwicky (1898–1974) discovered the phenomenon of "missing mass." Given the fact that its member galaxies are moving about in the cluster at hundreds of miles per second, the cluster should have disintegrated in less than a billion years—unless its total mass were sufficient to keep its constituent galaxies from flying apart. The cluster's total mass, he concluded, must be ten times greater than the combined visible star mass of its constituent galaxies. In other words, there had to be nine times more invisible or "dark" matter in the intergalactic medium than telescopes could reveal! Sixty years later, the nature of the "missing mass" remains one of the great unresolved issues facing today's astrophysicists. Dark matter is everywhere, from the smallest of dwarf galaxies to vast superclusters.

The fabric of superclusters is truly mind-boggling in that they form, not spheres, but sheet-like "pancakes" or long filamentary structures. "Pancakes" are roughly one-fifth as thick (40 million light-years) as they are wide; supercluster filaments can extend up to hundreds of millions of light-years across space.

Cosmic Voids

An even more startling discovery was yet to come. The universe contains vast regions tens of millions of light-

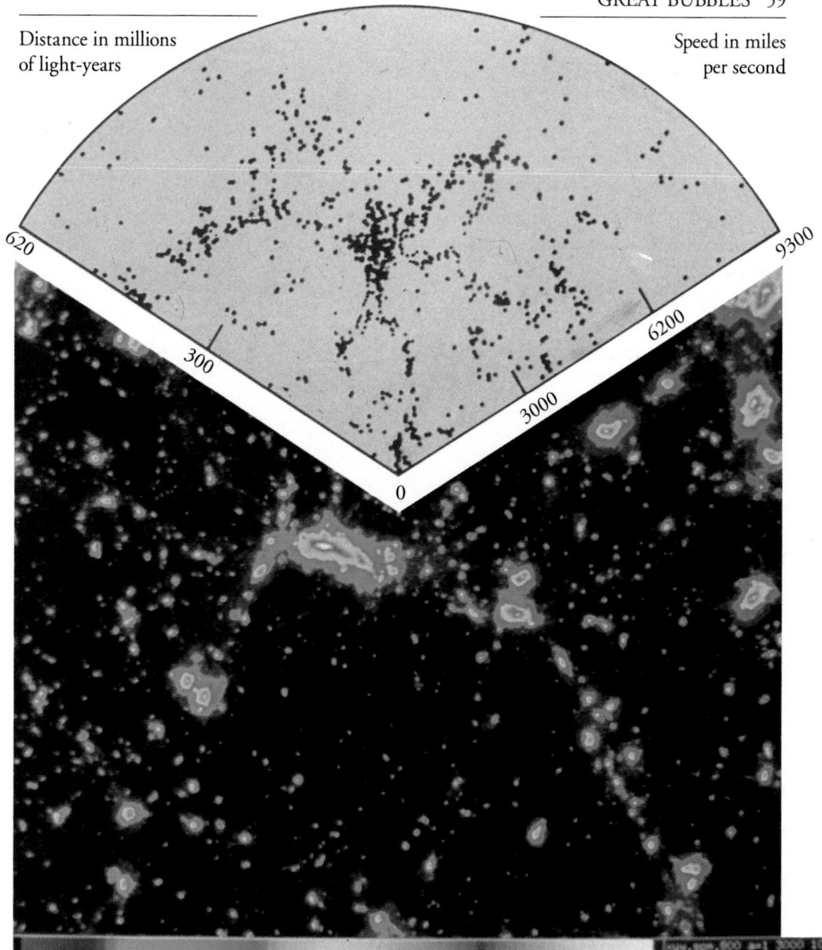

years across with no galaxies at all. Galaxies distributed in pancakes or filaments fill just one-tenth of the entire volume of the universe; the remaining nine-tenths are empty. Superclusters of both types appear to delineate one vast interconnected network of bubble-like voids. The galaxies are thus knitted into a fantastic patchwork, an immense tapestry of cosmic proportions, a skyscape of breathtaking scope and beauty.

Sky surveys indicate that galaxies are strung out in long filamentary structures around great voids (top). The same distribution takes shape in computer simulations of the universe (above).

The 20th century is the era of the Big Bang. The universe, most cosmologists now believe, was originally concentrated into an ultrasmall, ultrahot, ultradense point that exploded some 15 billion years ago. Newton's static cosmos has been turned into an expanding, restless space convulsed by violent events.

CHAPTER III
THE BIG BANG

Early in its history, the universe was like a huge atom smasher, creating and destroying elementary particles in unbelievably high-energy collisions. So that scientists can study these collisions in a particle accelerator, charged subatomic particles are sent into a bubble chamber where they interact with the atoms of liquefied gas, leaving behind a trail of tiny gas bubbles. The tracks of the particles (left) are curved by an intense magnetic field. One cartoonist's view of the time before the Big Bang, as expressed by an angel (right): "We're going to have to buy everything on the black market!"

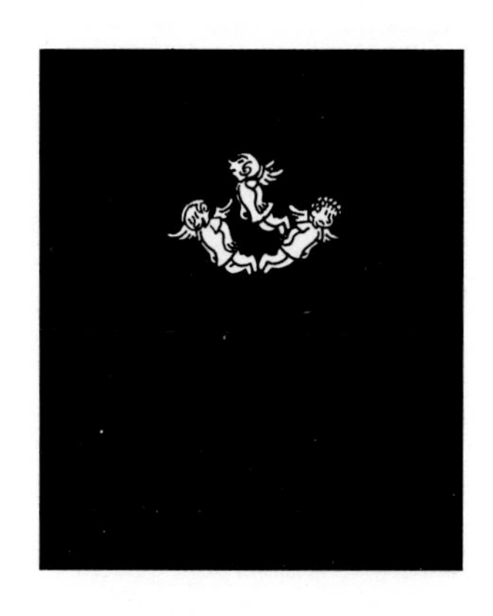

The Big Bang Theory: A Primordial Explosion

This turnabout in the way we view the cosmos began when American astronomer Edwin Hubble, who pioneered galaxy classification, discovered that the universe is expanding. In 1929 Hubble found evidence that distant galaxies were moving away from the Milky Way. Moreover, he discovered that their velocity was proportional to their distance: Galaxies twice as far apart were retreating from each other twice as fast; galaxies ten times farther apart, ten times faster.

Hubble also noticed that the expansion was the same regardless of the direction in which an observer chose to look. This meant that every galaxy had taken exactly the same amount of time to travel from its starting point to its current position.

Now let's run the film backward: About 15 billion years ago, all the galaxies in the universe were packed into a single point in space and time. A tremendous explosion, the Big Bang, must have been responsible for the expansion of the universe. With the Big Bang theory,

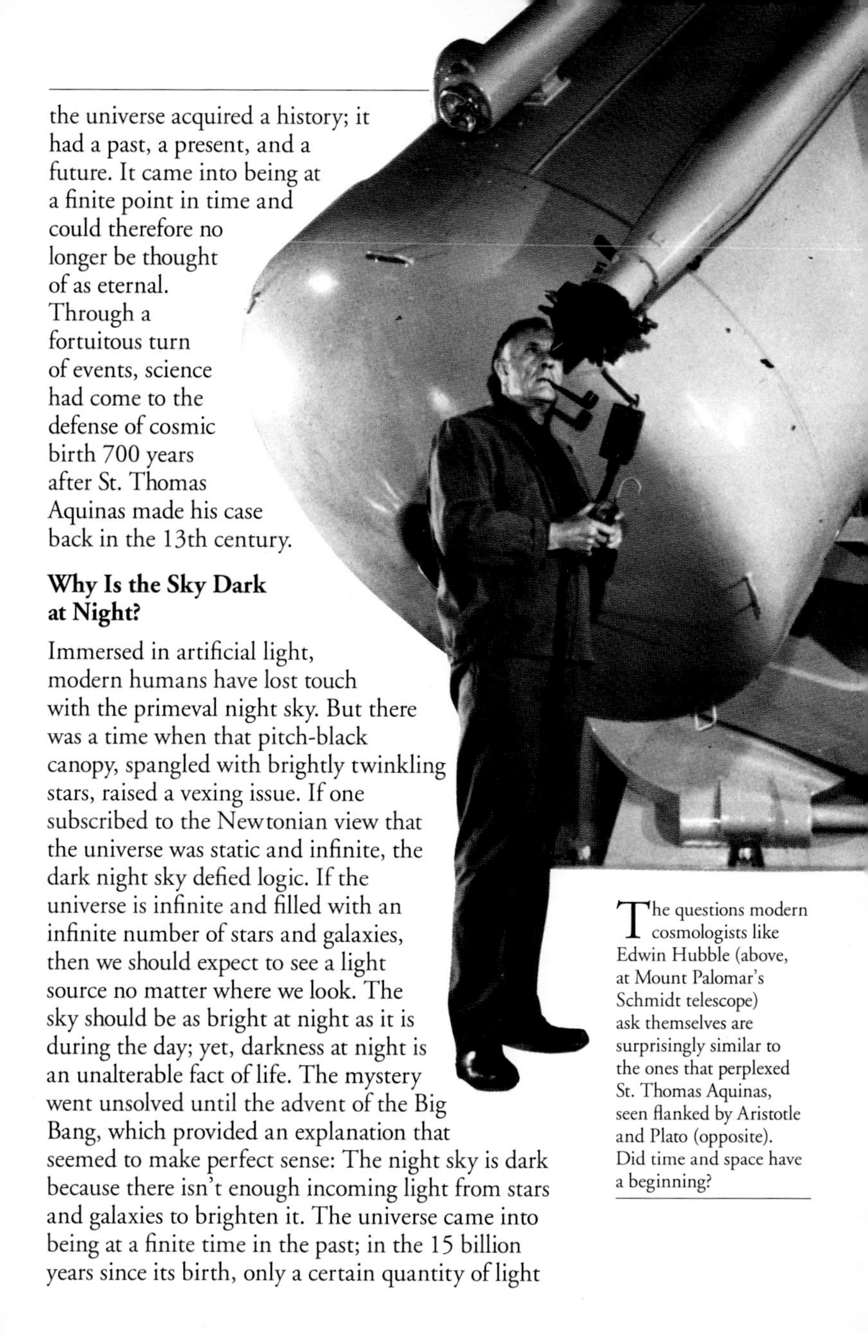

the universe acquired a history; it had a past, a present, and a future. It came into being at a finite point in time and could therefore no longer be thought of as eternal. Through a fortuitous turn of events, science had come to the defense of cosmic birth 700 years after St. Thomas Aquinas made his case back in the 13th century.

Why Is the Sky Dark at Night?

Immersed in artificial light, modern humans have lost touch with the primeval night sky. But there was a time when that pitch-black canopy, spangled with brightly twinkling stars, raised a vexing issue. If one subscribed to the Newtonian view that the universe was static and infinite, the dark night sky defied logic. If the universe is infinite and filled with an infinite number of stars and galaxies, then we should expect to see a light source no matter where we look. The sky should be as bright at night as it is during the day; yet, darkness at night is an unalterable fact of life. The mystery went unsolved until the advent of the Big Bang, which provided an explanation that seemed to make perfect sense: The night sky is dark because there isn't enough incoming light from stars and galaxies to brighten it. The universe came into being at a finite time in the past; in the 15 billion years since its birth, only a certain quantity of light

The questions modern cosmologists like Edwin Hubble (above, at Mount Palomar's Schmidt telescope) ask themselves are surprisingly similar to the ones that perplexed St. Thomas Aquinas, seen flanked by Aristotle and Plato (opposite). Did time and space have a beginning?

from stars and galaxies has had time to reach us. What's more, there are only a certain number of stars because they, too, have a finite lifespan. It may take millions or even billions of years, but eventually they do burn out.

Galaxies Are Moving Apart from One Another Within a Steadily Expanding Space

If all galaxies are moving away from us, wouldn't that put the Milky Way at the center of the universe? Actually, the fact is that it would appear to observers in any galaxy that all other galaxies are receding from their galaxy. Since every place is the center of the expansion, no place is the center.

To make this cosmic sleight-of-hand a bit more comprehensible, picture yourself inflating a balloon that is decorated with paper stars. As the surface of the expanding balloon stretches out, all the stars move farther and farther apart. The galaxies are set in space the way the paper stars are attached to the surface of the

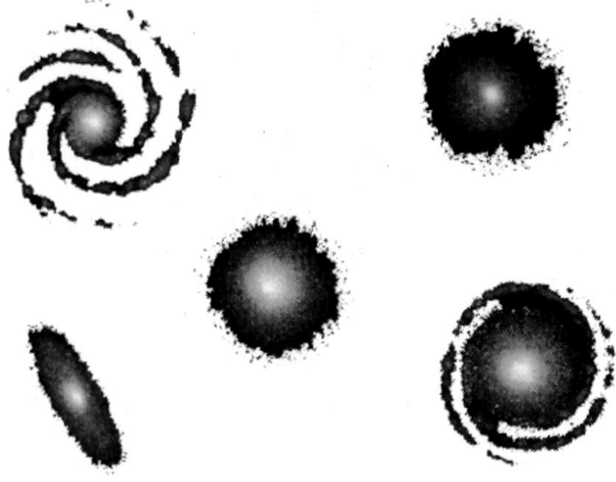

The surface of an inflating balloon (above) and the dispersal of galaxies (left and right) help us visualize the expanding universe. Will it expand indefinitely? Or will the attractive force of gravitation eventually overcome the initial outward thrust, start pulling galaxies together, and trigger a collapse culminating in an unimaginable implosion of energy and radiation? Will the Big Bang reverse itself into the Big Crunch?

balloon. Space is expanding, just as the surface of the balloon is the source of all motion. Any one of the paper stars would see all the others receding; the farther apart they are, the faster they appear to be moving away from each other. By the same token, the more distant a galaxy, the greater its relative speed of recession. The Big Bang theory turned the static Newtonian universe into dynamic space.

According to this model, galaxies are not drifting about within motionless space; space itself is expanding, and the galaxies with it: The space the universe takes up has been getting bigger with time. Fifteen billion years into cosmic history, the distance between any two galaxies has on average increased a thousandfold. The galaxies are not receding from the Milky Way. They are all receding from one another.

Continuous Creation in a Universe Without Beginning or End

Did the Big Bang theory take hold simply because the universe was found to be expanding? Not by a long shot, for astrophysicists are a cautious breed. In the 1950s a debate arose between proponents of the Big Bang scenario and advocates of the Steady State model, who took issue with the concepts of creation, evolution, and change inherent in Big Bang cosmology. They contended that the universe has had the same appearance at all times—and in the process gave Aristotle's unchanging heavens a new lease on life! Quite a few cosmologists found this model appealing because it avoided a creation event and its religious implications. But how could they reconcile a universe that is said to be changeless in time with its observed expansion? If the galaxies are steadily moving apart from one another, more and more vacant space must be left between them. To keep the universe looking eternally the same, these cosmologists found it necessary to postulate that exactly enough new matter is

One way to predict the future of the universe is to calculate its density. If it averages fewer than three atoms of hydrogen per cubic meter of space, the universe will keep expanding forever. If it averages more than three atoms, the cosmos will one day collapse on itself. But tallying every last particle of cosmic matter is problematic because the universe is known to contain an abundance of "dark matter." The total visible mass of stars and galaxies accounts for only a hundredth of the density needed to bring the expansion of the universe to a halt. Astronomers

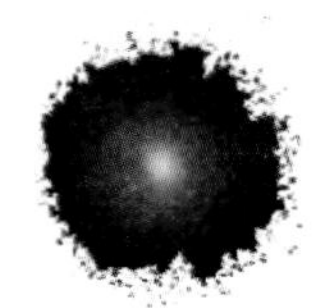

know that galaxies and galaxy clusters harbor an amount of invisible, or "dark," matter equal to ten times the visible mass. That would raise the average density of matter in the universe to ten percent of the critical value—not enough to reverse expansion.

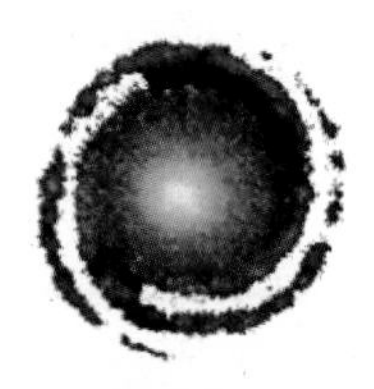

continuously created to fill in the space left empty by expansion. The rate of matter creation, they added, is so minute as to be virtually undetectable: About a single atom of hydrogen per quart of space per billion years would suffice. Intent on dodging a single creation event, Steady State theorists had to resort to an infinite series of mini-creation events.

The Afterglow of the Primordial Fireball

Paleontologists comb the African hinterlands for human fossils in hopes of reconstructing the early history of our species. Geologists probe the depths of the earth's crust in search of evidence that may prove helpful in reconstructing the early history of our planet. In much the same way, inquisitive astronomers scan the heavens for cosmic fossils that may hold clues to the early history of the universe. Because light does not travel instantaneously, looking out into space is the same as looking back into the past. Telescopes are time machines.

The discovery that the entire universe is bathed in relic radiation surviving from an early stage in its history, when it was just 300,000 years old, won over most scientists to the Big Bang theory and helped to discredit all alternative models. American physicist George Gamow (1904–68) postulated the existence of background radiation in 1946. Since an expanding universe would be expected to cool down and thin out as time wore on, Gamow reasoned, it must have once been much hotter and denser than it is now. The relationship between the two components of the universe—matter (atoms, humans, stars, galaxies) and radiation—must

Proponents of the Steady State theory—British astronomers Herman Bondi, Thomas Gold, and Fred Hoyle—postulated that new galaxies are continuously created. Otherwise, the average density of galaxies in the ever-expanding space of the universe could not remain constant. There was, they argued, no change in the average distance between galaxies (a principle illustrated below). According to the Big Bang theory, the space left empty by receding galaxies is not counterbalanced by newly created ones; it just keeps getting bigger and bigger, at least in a cosmos that expands forever.

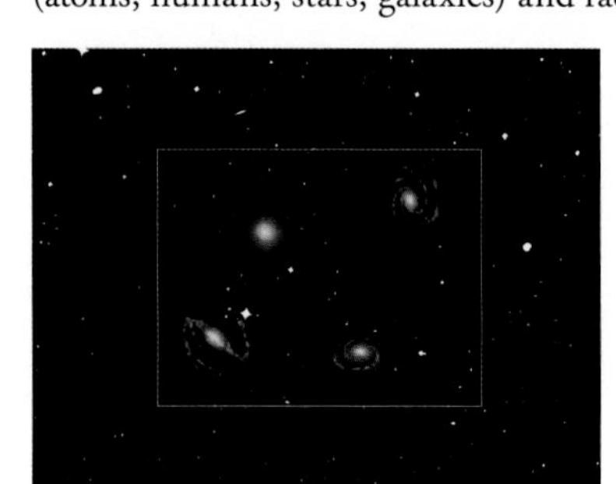

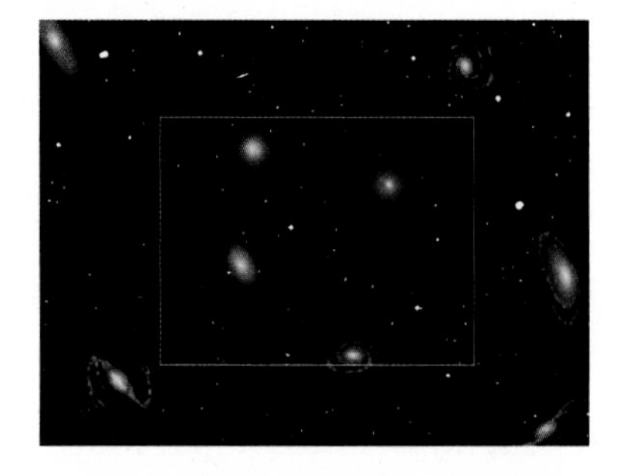

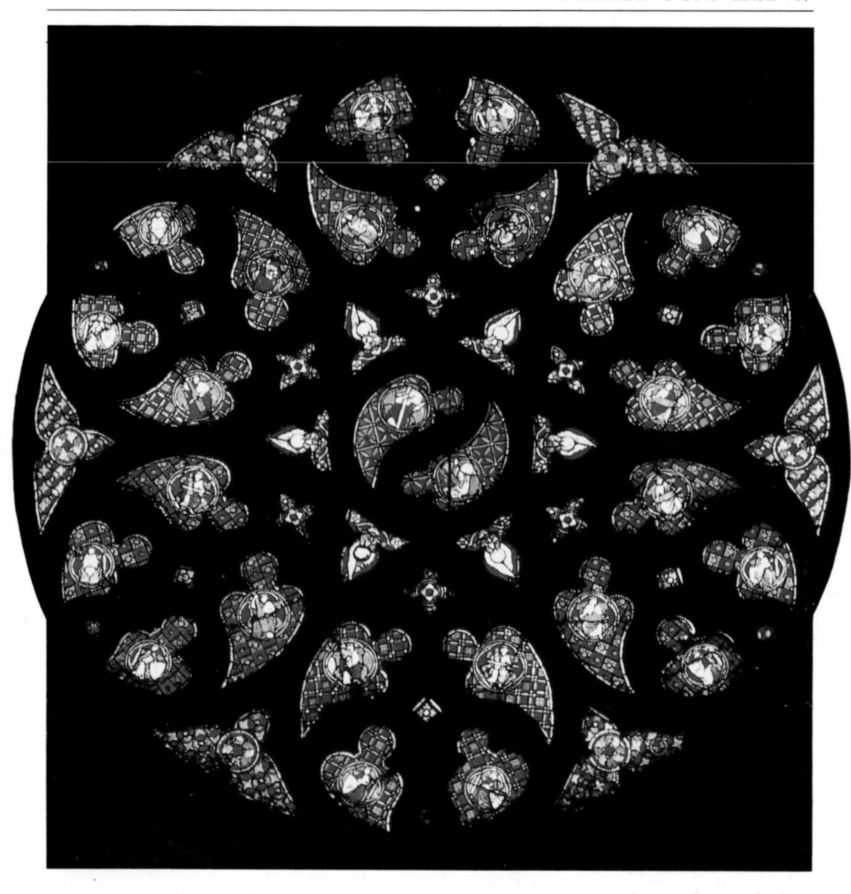

have been reversed when the universe was younger. According to Albert Einstein (1879–1955), mass is a form of energy. Matter dominates the universe as we know it; its energy density is about three thousand times greater than the energy density of radiation. Very early on, however, between 1 second and 300,000 years after its birth, the universe was dominated by radiation. These hot, high-energy rays were set free 300,000 years after the Big Bang; by then their temperature had dipped to 10,000 degrees Kelvin. (Kelvin degrees are equivalent to Celsius degrees, but 0 degrees K is 273 degrees lower.)

An analogy to the Big Bang theory can be found in this stained-glass window in Saint-Bonaventure cathedral in Lyons, France. There is no center, and the constituent designs seem to be moving away from one another much as galaxies are receding in the expanding space of the universe.

We should, Gamow predicted, still be able to detect them today. Over the intervening 15 billion years it has taken them to reach the Milky Way, one would expect them to have lost energy, which would translate into a drop in temperature. By now, he said, these observable relics should have cooled all the way down to about 3 degrees K.

Such cosmic fossils would be the afterglow of the primordial fireball, comparable to the lingering light

With fellow Americans Ralph Alpher and Robert Herman, George Gamow (left) was the first to postulate a primeval fireball and relic radiation. Substantially cooled by cosmic expansion, that fossil radiation could be detected only with a radio telescope. When Arno Penzias and Robert Wilson accidently picked up radiation at 3 degrees K, they suspected at first that it was caused by a pair of pigeons nesting in their telescope (below).

and warmth radiated by embers in a fireplace. Over the next 20 years, no one took the trouble to investigate these remnants of creation. Physicists had unconscious misgivings about the religious implications of Big Bang cosmology, and Gamow's prediction was forgotten. Then, in 1965, Arno Penzias and Robert Wilson, American radio astronomers working for Bell Laboratories, accidently tuned in to the "music" of creation with an ultrasensitive radio antenna designed to track Telstar, the first communications satellite.

The expansion of the universe is one of the observational mainstays of the Big Bang model. Another is the chemical composition of the universe: Three-quarters of its mass is in the form of hydrogen, the rest mostly in the form of helium. Still another is the cosmic background radiation, emitted primarily at microwave wavelengths. However, since only a little of it leaks through the earth's atmosphere, it cannot be measured wholly from the ground. The mission of the Cosmic Background Explorer (COBE) satellite (above) launched by NASA in 1989 is to gather data about the microwave background. Its findings are consistent with a universe originally squeezed into an ultrahot, ultradense speck of space.

The background radiation they had chanced upon proved to be the undoing, not just of the Steady State model, but of all alternatives to the Big Bang scenario.

All Matter in the Universe "Froze Out" of an Energized Void

Our story begins a split-second after the primordial explosion, at precisely

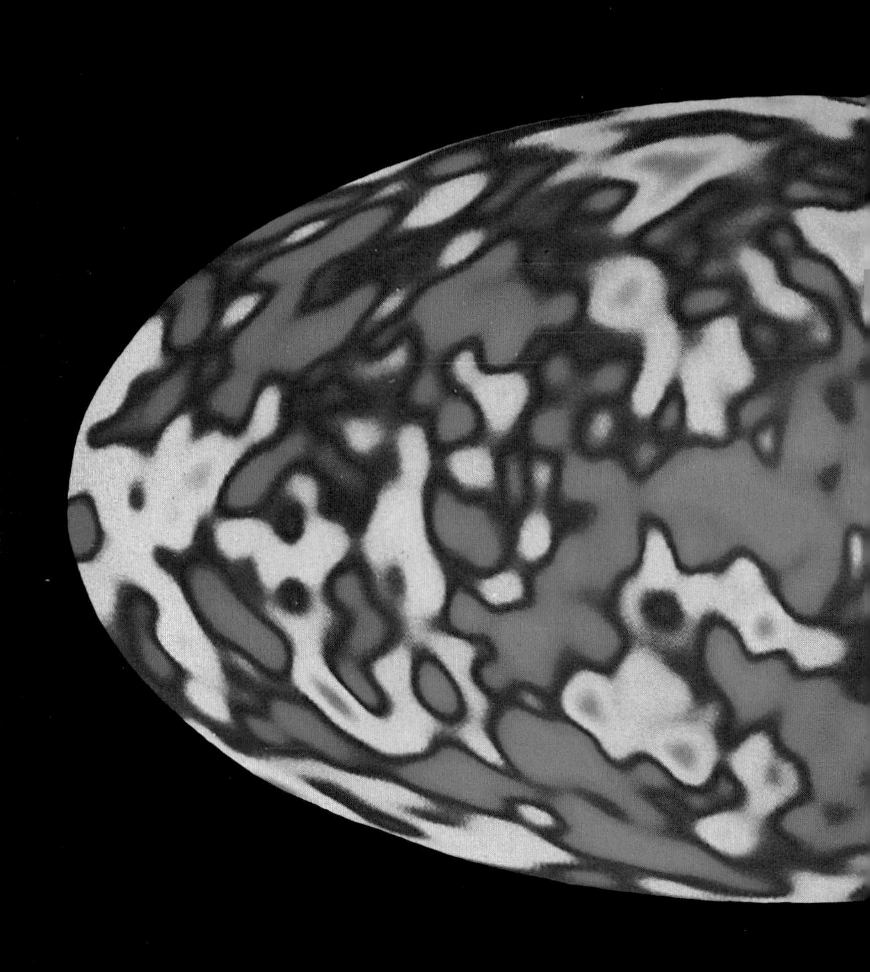

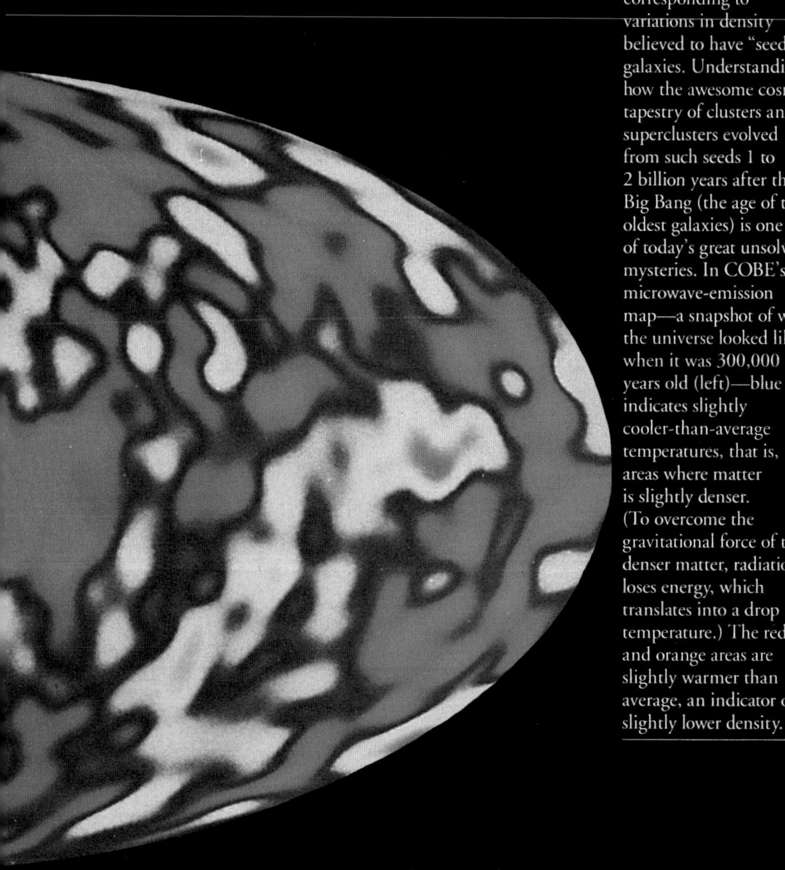

COBE has detected minute temperature fluctuations of about 30 millionths of a degree in the relic radiation, corresponding to variations in density believed to have "seeded" galaxies. Understanding how the awesome cosmic tapestry of clusters and superclusters evolved from such seeds 1 to 2 billion years after the Big Bang (the age of the oldest galaxies) is one of today's great unsolved mysteries. In COBE's microwave-emission map—a snapshot of what the universe looked like when it was 300,000 years old (left)—blue indicates slightly cooler-than-average temperatures, that is, areas where matter is slightly denser. (To overcome the gravitational force of the denser matter, radiation loses energy, which translates into a drop in temperature.) The red and orange areas are slightly warmer than average, an indicator of slightly lower density.

The titanic explosion that spawned the universe, as interpreted by artist Franck Malina (left). American physicist Alan Guth describes the ensuing surge of expansion as the "inflationary epoch." In a split-second (10^{-35} to 10^{-32} second), the observable universe swelled from a speck much tinier than the nucleus of an atom to the size of an orange.

10^{-43} second (a decimal point followed by 42 zeros and a 1). What happened before that? No one knows. The universe had a temperature of 10^{32} degrees K (1 followed by 32 zeros)—an inferno that would have strained even Dante's imagination—and filled a spherical pinpoint of space one three-thousandth of an inch in diameter.

There were no atoms yet, let alone stars or galaxies: All was empty. We tend to think of it as a serene, hushed nothingness, an insubstantial, uneventful void, but it was in fact seething with all the pent-up energy of the primordial explosion.

The cosmic clock struck 10^{-32} second. Expansion left the universe slightly thinned out and slightly cooler. The first elementary particles made their appearance. The resulting "soup" of quarks (building blocks of matter), electrons (electricity particles), and neutrinos (electrically neutral particles with little or no mass) was awash with photons (particles of light).

When matter came into being, a mirror image, antimatter, was created, which counterbalanced its electrical charges and thus preserved the electric neutrality of the universe. Matter and radiation were constantly interacting. When a particle and its antiparticle collided, mass was converted into the energy of photons. Photons were in turn converted back into particle-antiparticle pairs. Matter, antimatter,

Quark
Antiquark
Quark
Antiquark
Positron
Electron
Neutrino
Antineutrino
Photon
Direction of particles

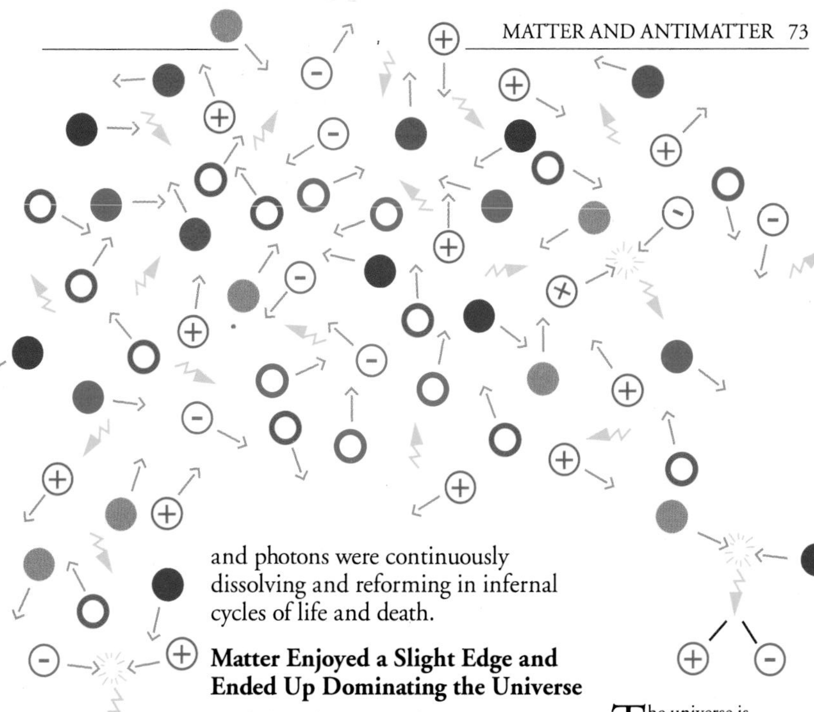

and photons were continuously dissolving and reforming in infernal cycles of life and death.

Matter Enjoyed a Slight Edge and Ended Up Dominating the Universe

Had there been exactly as many antiparticles as there were particles, our story would end there. Matter and antimatter would have annihilated each other, leaving only a radiation-filled universe devoid of elementary particles, stars, and galaxies, much less human beings. Fortunately for us, however, nature did not treat matter and antimatter equitably; matter enjoyed a very slight edge. There were a billion and one matter particles for every billion antimatter particles that emerged from the void. For every billion particle-antiparticle pairs that mutually annihilated and converted into a billion photons, a single particle of matter survived.

Increasingly Complex Structures Took Shape as the Universe Cooled Down and Thinned Out

Things shifted into high gear when the cosmic clock struck a millionth (10^{-6}) of a second. The universe had

The universe is governed by four fundamental forces. Gravitation holds the planets in their orbits around the sun and binds stars into galaxies. The electromagnetic force binds electrons in atoms and is the agent of chemical bonds, assembling molecules into long chains of DNA. The two nuclear forces control the world of atoms. The weak force is responsible for radioactive decay, while the strong force glues protons and neutrons together to form atomic nuclei.

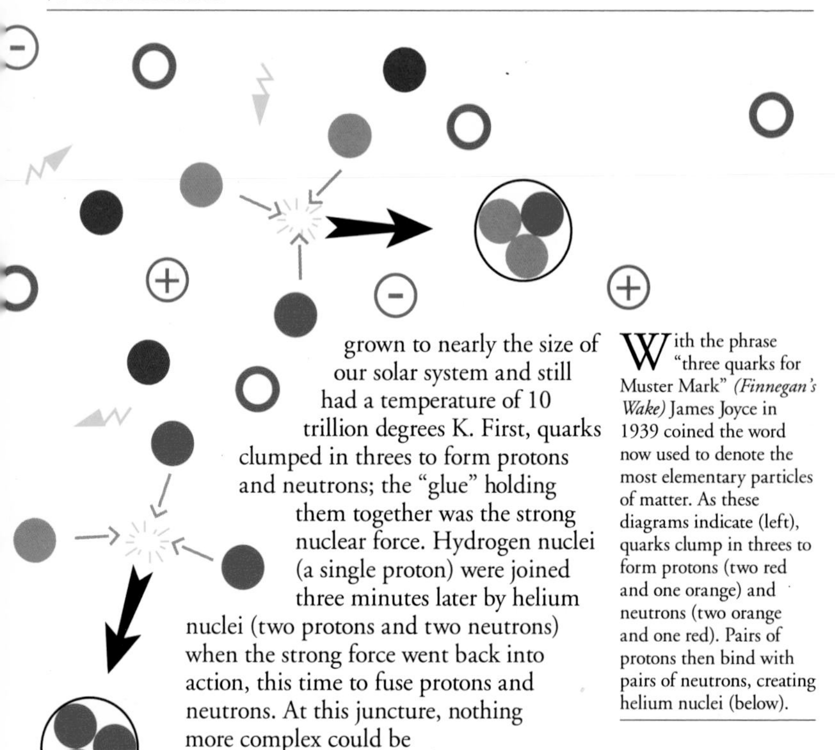

grown to nearly the size of our solar system and still had a temperature of 10 trillion degrees K. First, quarks clumped in threes to form protons and neutrons; the "glue" holding them together was the strong nuclear force. Hydrogen nuclei (a single proton) were joined three minutes later by helium nuclei (two protons and two neutrons) when the strong force went back into action, this time to fuse protons and neutrons. At this juncture, nothing more complex could be processed: Helium nuclei were spreading out through the expanding space of the universe, all but eliminating any chance of further nuclear collision and clumping.

With the phrase "three quarks for Muster Mark" *(Finnegan's Wake)* James Joyce in 1939 coined the word now used to denote the most elementary particles of matter. As these diagrams indicate (left), quarks clump in threes to form protons (two red and one orange) and neutrons (two orange and one red). Pairs of protons then bind with pairs of neutrons, creating helium nuclei (below).

Awash with Hydrogen Nuclei, Helium Nuclei, Electrons, Photons, and Neutrinos, the Universe Kept on Expanding

The next 300,000 years were relatively uneventful. The universe cooled to 10,000 degrees K. Full-fledged atoms now appeared as electromagnetic force joined solitary electrons to single-proton hydrogen nuclei, creating hydrogen atoms, and two electrons to helium nuclei, forming helium atoms.

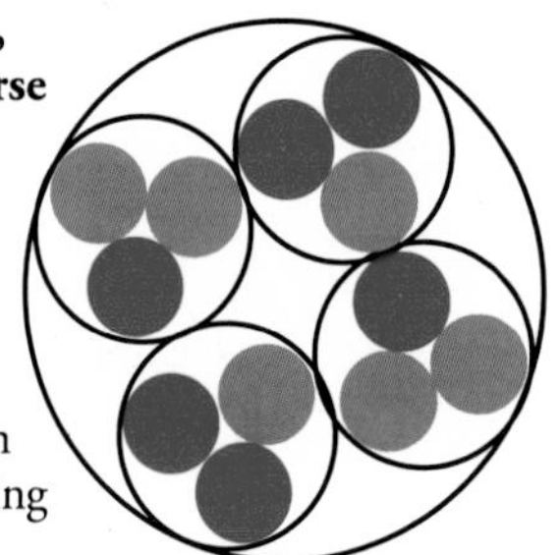

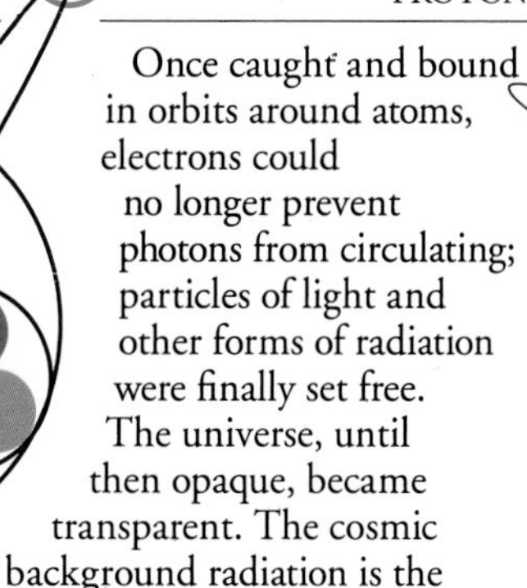

Once caught and bound in orbits around atoms, electrons could no longer prevent photons from circulating; particles of light and other forms of radiation were finally set free. The universe, until then opaque, became transparent. The cosmic background radiation is the cooled residue of the photons that burst forth during this stage.

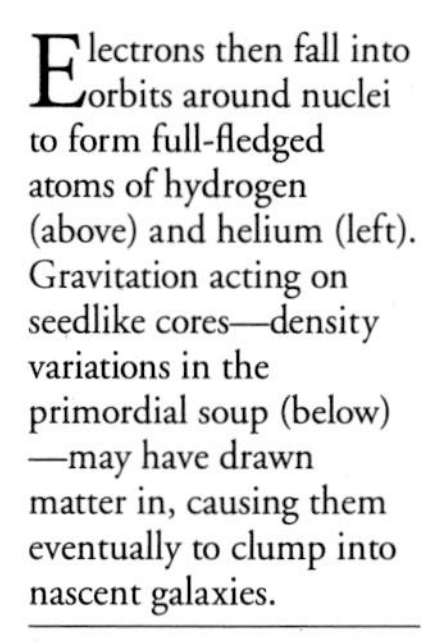

Electrons then fall into orbits around nuclei to form full-fledged atoms of hydrogen (above) and helium (left). Gravitation acting on seedlike cores—density variations in the primordial soup (below)—may have drawn matter in, causing them eventually to clump into nascent galaxies.

We last saw the universe at a stage when it was unable to synthesize anything more complex than helium. But now that complete hydrogen and helium atoms were available, all that was about to change. Abetted by gravity, "oases" of heat started to emerge in the frigid emptiness of space—oases that eventually became galaxies. Their constituent matter, held together by gravitational attraction and therefore unaffected by the expansion of the universe, escaped the cooling and diluting that had until then prevented matter from progressing to more complex structures.

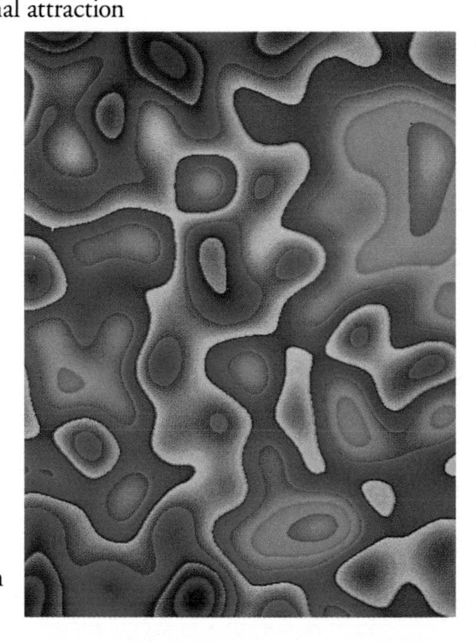

These oases of heat, however, had one major drawback: The particles within them were too thinly spread out. With an average density of just one atom of hydrogen per cubic centimeter, nascent galaxies were millions of billions of times more rarefied than the air we breathe. A denser medium would be needed to encourage atomic collisions.

Then, inside galaxies, stars began to form.

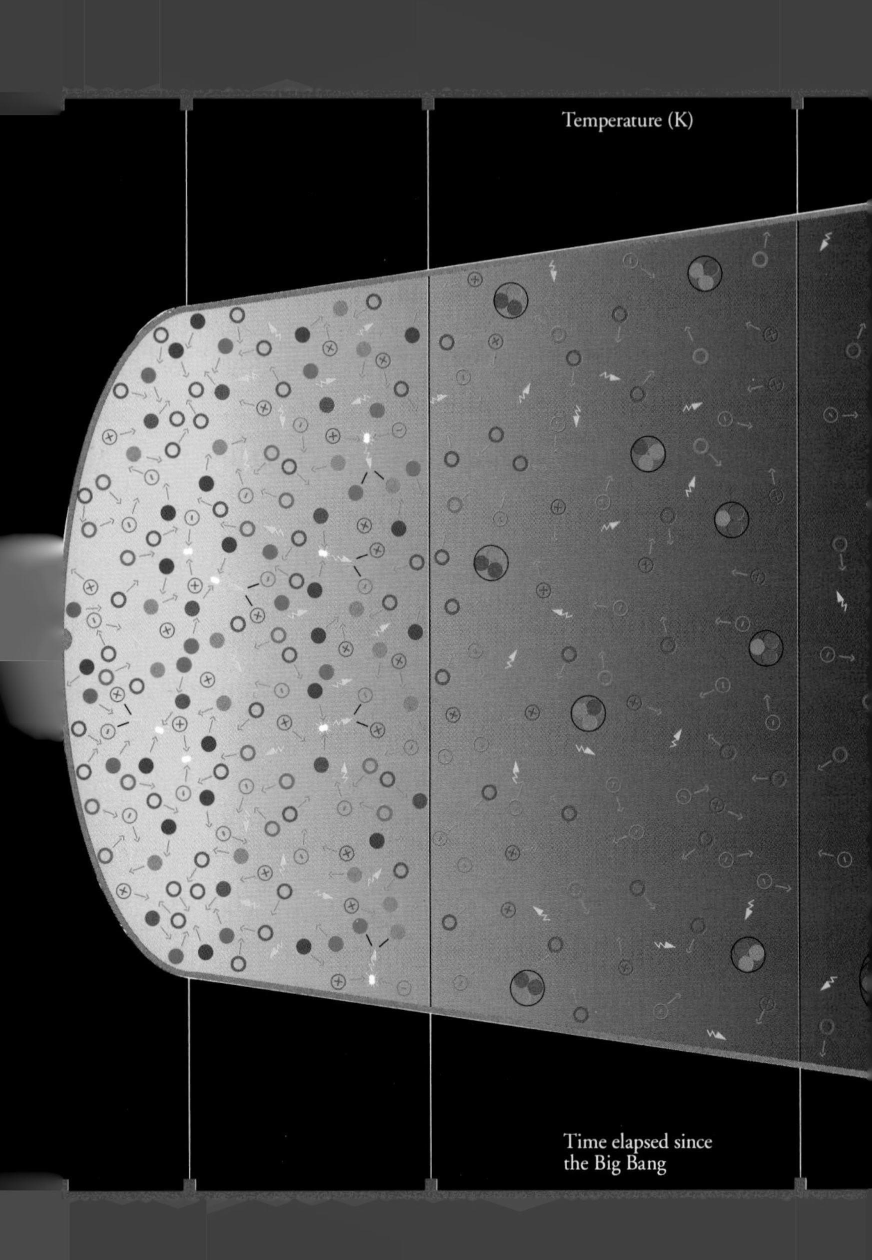
Temperature (K)
Time elapsed since the Big Bang

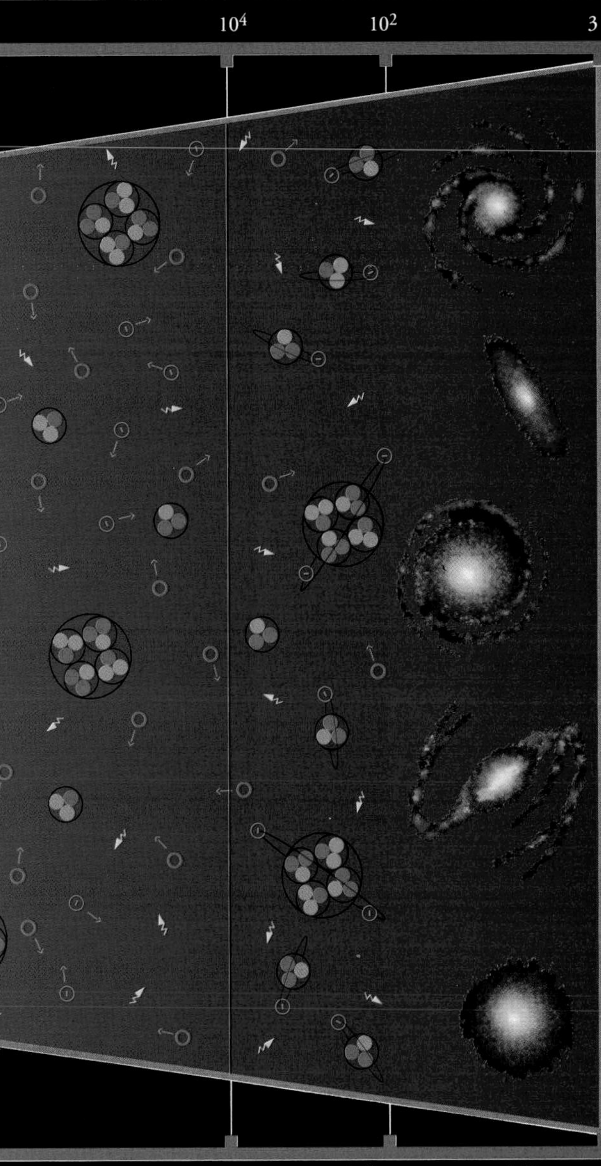

Recent findings in elementary particle physics have allowed astrophysicists to outline the history of the universe all the way back to 10^{-43} second after the Big Bang. They do not know enough at the present time to speculate beyond that point. During the split-second-long "inflationary epoch," a soup of quarks, electrons, neutrinos, photons, and their antiparticles "froze out" of the energized void. Neutrons and protons—components of all future atomic nuclei—came into being at 10^{-6} second. By the third minute, with the emergence of stable hydrogen and helium nuclei, 98 percent of the mass of the universe had taken shape as hydrogen (75 percent) and helium (23 percent). Three hundred thousand years later, electrons combined with hydrogen and helium nuclei, creating hydrogen and helium atoms. From then on, the universe had the raw material it needed to make stars and galaxies.

Just as the universe changes, contrary to what Aristotle had supposed, so do the stars in it. The story of stellar birth, life, and death is very much bound up with our own, for in the final analysis we humans are nothing but stardust.

CHAPTER IV

THE LIFE AND DEATH OF STARS

Hundreds of young stars emitting intense ultraviolet radiation impart a lovely glow to stellar nursery IC 1283-4 in the constellation Sagittarius (opposite). The remnant of the supernova, or exploding star, Tycho Brahe observed in 1572 (right) is an intense source of radio waves produced as electrons are flung into space at ultrahigh speed.

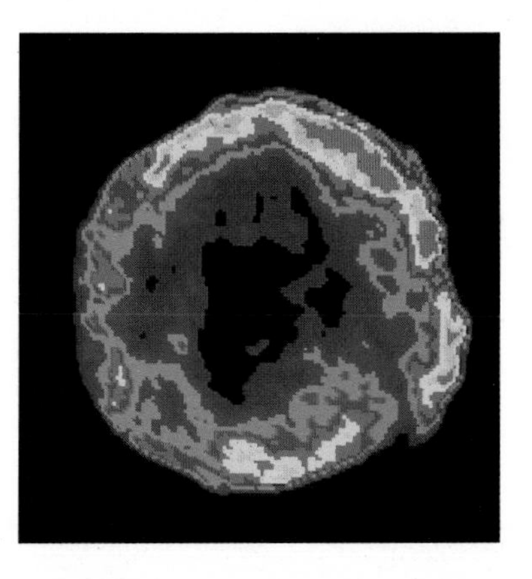

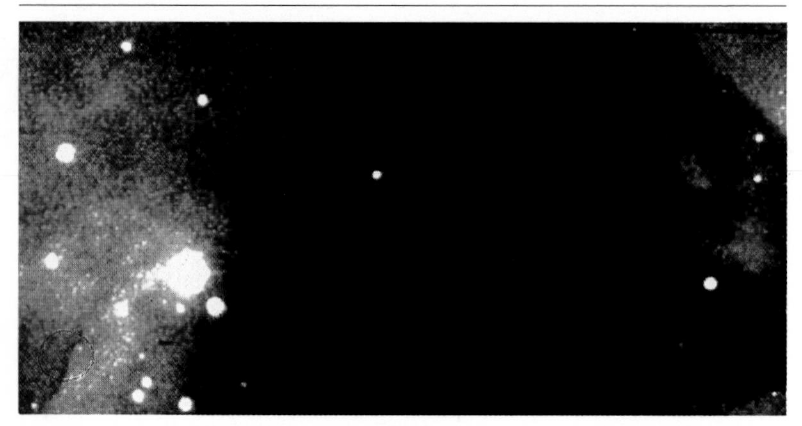

Heat from Fusion Turned on the Protostars: Baby Stars Filled Stellar Nurseries

How can we penetrate stellar nurseries and witness the birth of stars firsthand when the visible light young stars emit is absorbed by the gas and dust surrounding them? In an optical photograph (above), the center of stellar nursery NGC 2024 in the constellation Orion is a starless dark patch in the sky. Pinpoints of light are visible only on the periphery of the star-forming region, where light is not obscured by dust particles.

The first stars came into being when the universe was about 2 billion years old. Gravity caused nascent galaxies to collapse and fragment into hundreds of billions of pockets of hydrogen and helium. These gas cloudlets took on a spherical shape as they, too, contracted under their own gravity. Gradually the matter in their cores became more tightly packed, until they were 160 times denser than water. The temperature of these central regions soared to tens of millions of degrees K. Forged only minutes after the Big Bang, hydrogen and helium atoms at the cores of the now-spherical gas cloudlets went wildly crashing into each other: Electrons, hydrogen nuclei (that is, lone protons), and helium nuclei were released from their atomic straitjackets.

This scenario recalls what took place when the universe was 3 minutes old. Ultrahigh heat and density allowed nature to indulge once again in its favorite activity: fusion. Protons clumped in fours, creating helium nuclei and in the process releasing energy in the form of radiation. Protostars "switched on" once they got hot enough to convert some of their proton mass into energy. A helium nucleus has slightly less mass than do its four individual constituent protons; the difference in mass is converted

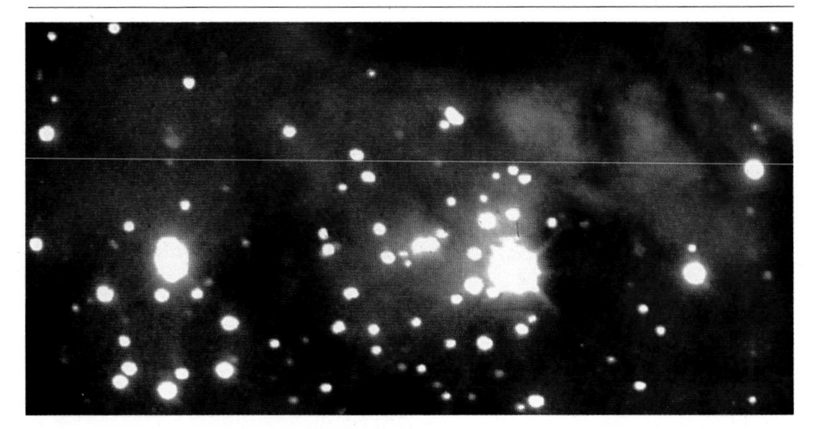

into energy. It is that energy that caused spherical gas condensations to ignite and achieve full-fledged stardom. Protogalaxies now functioned as gigantic stellar nurseries.

All of a sudden, with that release of energy, the contraction of gas spheroids came to an immediate halt. The compressive force of gravitation holding protostars together was exactly offset by the outward pressure of radiation that would have otherwise blown them apart.

Since infrared waves pass unabsorbed through dust particles, the development of infrared detectors in the 1980s unlocked at least some of the mystery of the birth of stars. An infrared image of NGC 2024 reveals a region teeming with hundreds of hot, brilliant young stars (above).

Dark dust lanes crisscrossing a stellar nursery delineate the "petals" of the Trifid Nebula (infrared image, left), set aglow by the hundreds of newborn stars at its center.

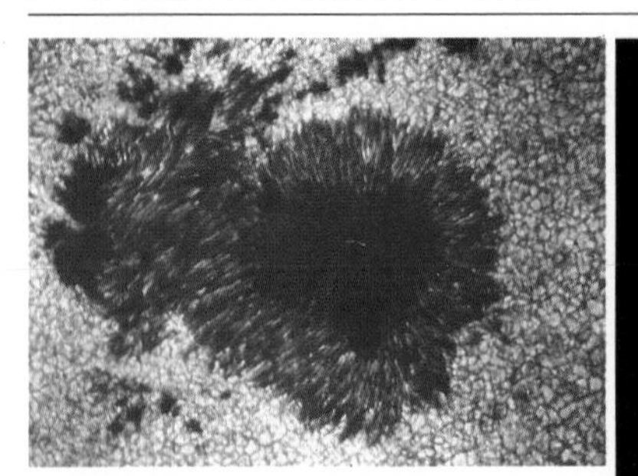

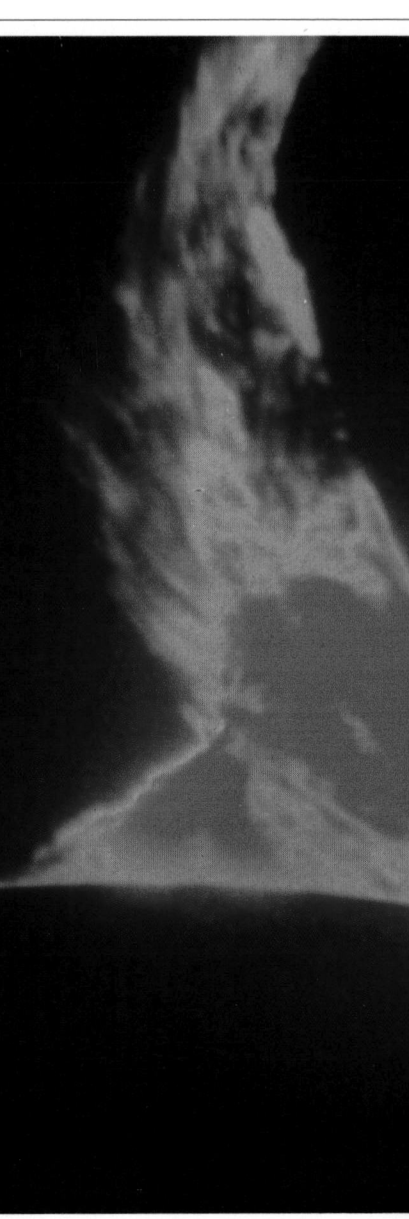

Portrait of a Star: Our Sun

The sun was born 4.6 billion years ago when a gas cloud in the Milky Way started to undergo gravitational collapse, possibly induced by the explosive death throes of a nearby star (supernova). It wandered out of its progenitor cloud of gas and dust and became a solitary source of light and energy, liberally shedding its warmth on the nine planets around it. Thanks to that solitary star, life eventually emerged on one of them, our own blue earth.

A closer look at our star reveals a truly spectacular sight. Baked to 6000 degrees K by the cauldron at the solar core, the shifting, fiery, gaseous surface of the sun is broken up into thousands of gigantic gas cells, each thousands of miles in diameter. These so-called granules rise and sink in a cycle of life and death that lasts just a few minutes. Sunspots, a phenomenon first seen by Galileo, dot the solar landscape. Up to thousands of miles across—the size of a small planet—sunspots look dark because they are some 2000 degrees K cooler than the surface.

The surface of the sun churns with intense activity; it must vent its energy somewhere. Every now and then an entire sunspot will blaze up, spurt tongues of flame, and shoot great bursts of matter into space. Some of these get caught up in magnetic fields, forming graceful loops of light as they arch back toward the surface. These fiery eruptions fling streams of protons and electrons into space, where they join the solar wind of charged particles flowing outward from the sun's topmost layers.

The Universe Was Given a Second Chance

The core of a star becomes rich in helium as it depletes its supply of hydrogen fuel. As the outward pressure exerted by radiation grows weaker, gravity gains the

The average number of sunspots (photograph, opposite left) increases and decreases in an 11-year cycle. This phenomenon, astrophysicists believe, can be correlated with the periodic rearrangement of magnetic fields in the sun's interior: Every 11 years, solar polarity reverses direction (north to south and vice versa).

The solar wind pushes out the tails of comets, those "dirty snowballs" that drop in on our solar system from time to time (Comet West, left). As the sun's heat evaporates surface ices, this stream of charged particles blows them back into a straight tail that always points away from the sun. During periods of peak solar activity (photograph, center), solar wind gusts disrupt radio communications on earth.

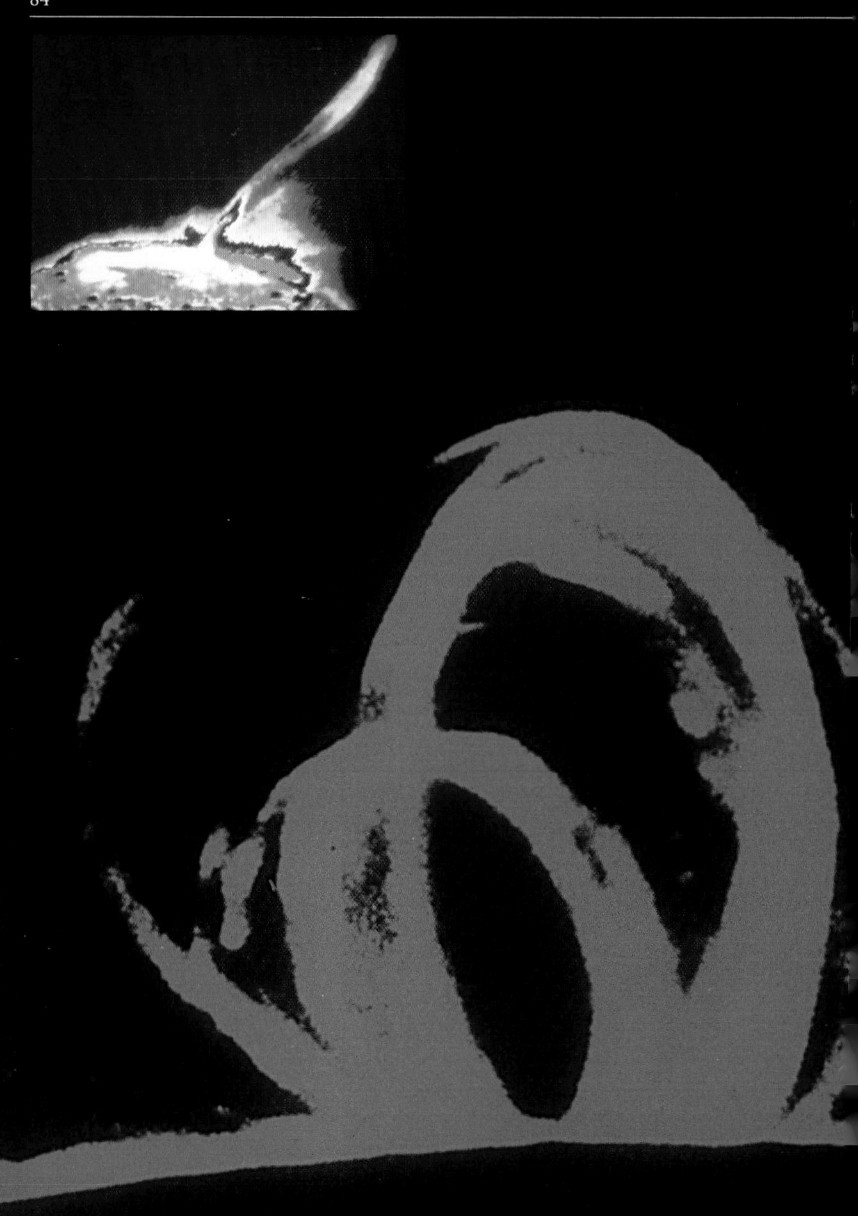

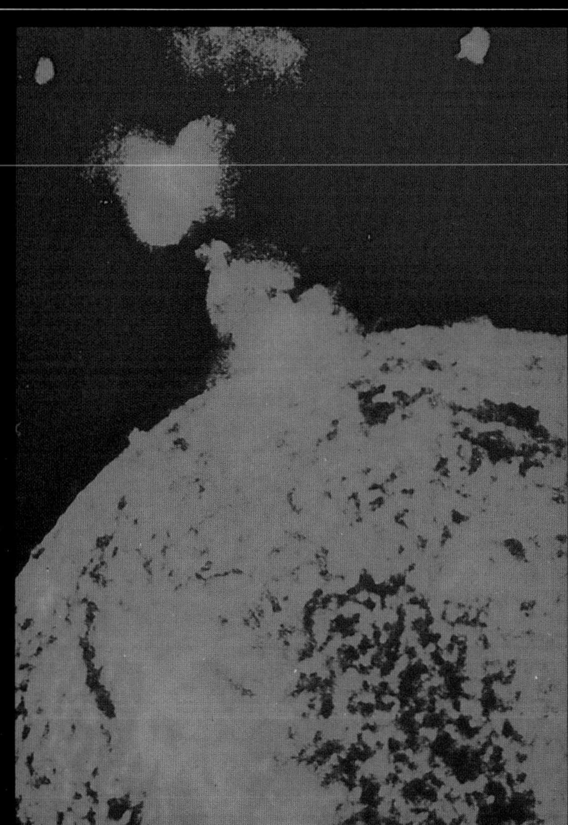

Solar Outbursts

Were it not for the sun, we would have no idea what a stellar surface looks like up close; all other stars are simply too far away to be studied. The spectacular solar eruptions captured in these three ultraviolet images are flamelike prominences leaping thousands of miles above the surface. Some prominences stream away into space as protons and electrons traveling at about 600 miles per second; others arch back toward the surface, forming graceful loops. Solar flares are thought to be associated with magnetic phenomena. Because the sun is gaseous, its period of rotation is longer at the poles (35 days) than at its equator (25 days). Differential rotation twists and entangles lines of magnetic force in the solar interior. These eventually break and pop out through the surface, producing sunspots and prominences.

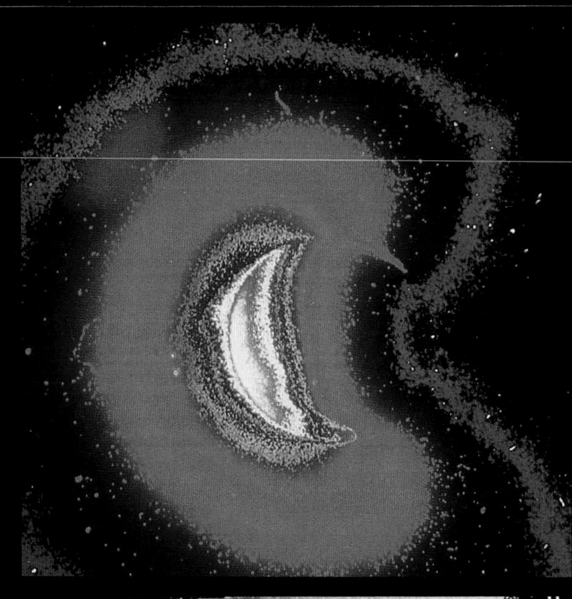

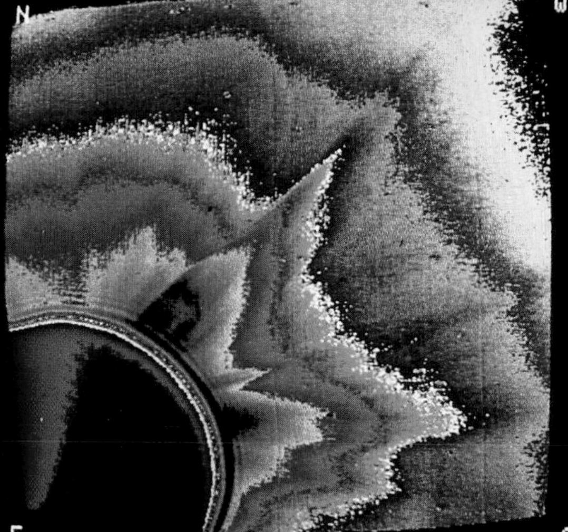

The Solar Corona

In 1973 the *Skylab* telescope obtained an image (opposite) revealing the effect of a gigantic solar eruption on the solar corona (violet area), which has a temperature of millions of degrees and extends millions of miles into space. NASA's *Solar Maximum Mission* spacecraft, designed to study the sun during periods of peak activity, afforded scientists an even more detailed view of the solar corona (left below). The colors vary with density, violet indicating maximum density and yellow minimum density. Even in areas of highest electron concentration, the solar corona is a near vacuum. A photo taken by an *Apollo 16* astronaut (left above) shows the effect of the solar wind on the tenuous envelope of hydrogen around the earth, creating a halolike aura detectable only at ultraviolet wavelengths.

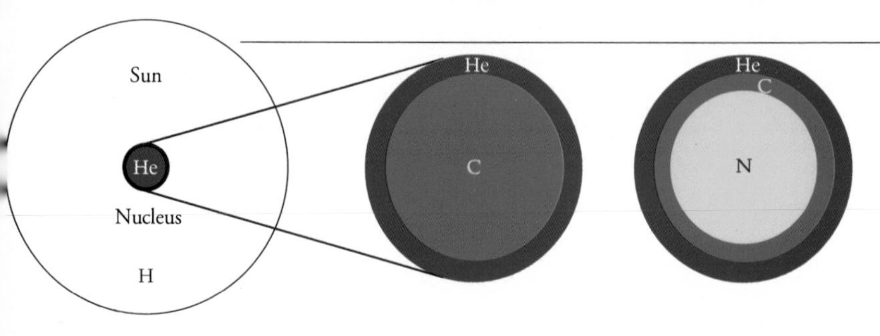

upper hand. The helium core of the star begins to contract, increasing its density and temperature as well as that of the hydrogen envelope surrounding it. Eventually the outer layers reach temperatures in excess of 10 million degrees K, triggering another round of hydrogen fusion reactions. The renewed "burning" of hydrogen releases tremendous surges of energy, swelling the star to a hundred times its original size and causing its light to redden. It is now a red giant.

But the hydrogen stored in its outer layers cannot last indefinitely; in due course it, too, must run out. The helium core shrinks still more for lack of fuel, heating the center to over 100 million degrees K and triggering the burning of helium. Three helium nuclei are fused to make a carbon nucleus. How does nature work this wonder? A carbon nucleus has slightly less mass than its three constituent helium nuclei. The difference in mass is converted into energy.

Why was the helium barrier broken inside stars and not in the Big Bang? Fusing three helium nuclei required a tremendous concentration of matter and time, which the expanding universe did not have to spare. Matter was spreading out, inexorably, as time wore on; the odds of nuclear fusion were already virtually nil by the third minute after the Big Bang. But red giants were unaffected by the fact that space was expanding and matter diluting. They had billions of years to help thermonuclear reactions along. Stars became cosmic crucibles, forging the chemical elements without which life would not have been possible. They rescued the universe from sterility.

The temperature of a red giant is not uniform; it ranges from billions of degrees K at the core to thousands of degrees K at the surface. Temperatures of 10 million degrees are needed for the fusion reactions that convert hydrogen into helium, 100 million degrees to initiate the conversion of helium into carbon, and so on. As the site of nucleosynthesis moves outward, elements are laid down in successive layers, leaving the star with an "onionskin" structure of concentric shells made of progressively lighter chemical elements (diagram, above).

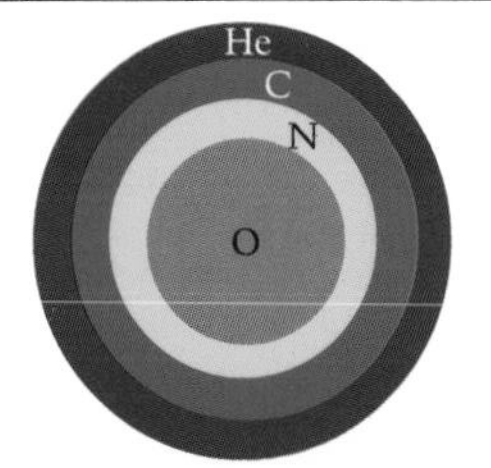

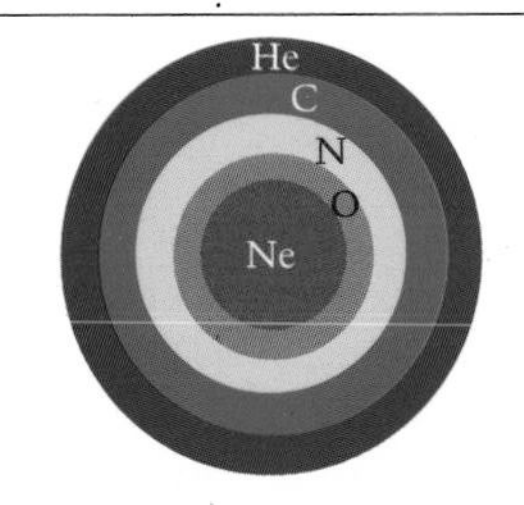

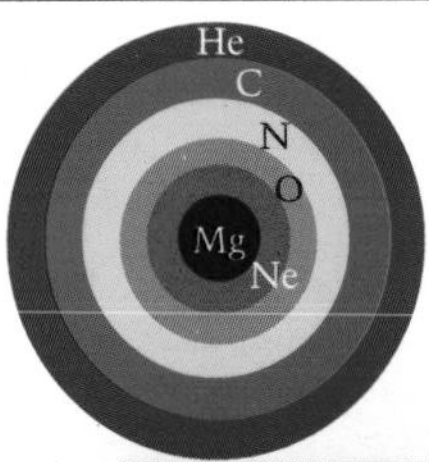

Recalcitrant Iron

The creative alchemy at work in stars produced more than twenty new chemical elements in just a few million years. Once their supply of helium ran out, they fused carbon to make oxygen; once the carbon was depleted, oxygen started to burn, and so on, building up to heavier elements such as neon, magnesium, aluminum, and sulfur.

By the time iron made its appearance, stellar cores had become repositories of the chemical elements responsible for the diversity of life on earth, the stuff destined to make up more than 90 percent of the atoms in our bodies. Iron, these starry masons' most impressive structure, has a nucleus of twenty-six protons and thirty neutrons.

But unlike the elements synthesized before it, iron cannot be used as nuclear fuel. Thermonuclear reactions involving iron use up energy, instead of producing it. Once it has exhausted its fuel, a star stops generating energy. With no outward radiation pressure to offset the inward pull of its own gravity, a star must collapse and die.

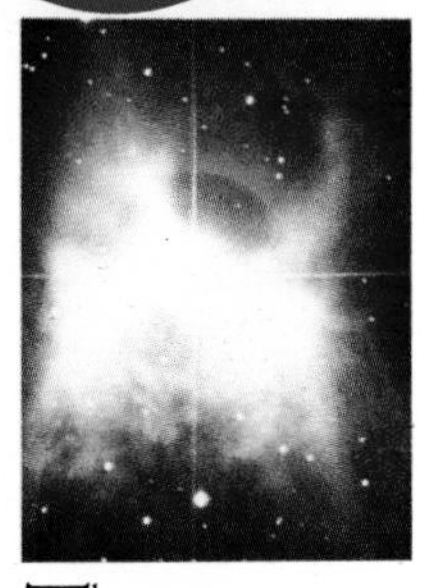

The energy thermonuclear reactions generate in the innermost quarter of the sun's radius radiates eight-tenths of the way toward the outer layers, where convection currents carry it to the surface (diagram, below left). The journey from the solar core through the radiation zone to the surface takes 100,000 years. Stars that have exhausted their core of hydrogen fuel balloon out to red giants like HD 65750 (above), seen here enshrouded in gas and dust resulting from the star shedding its outer layers.

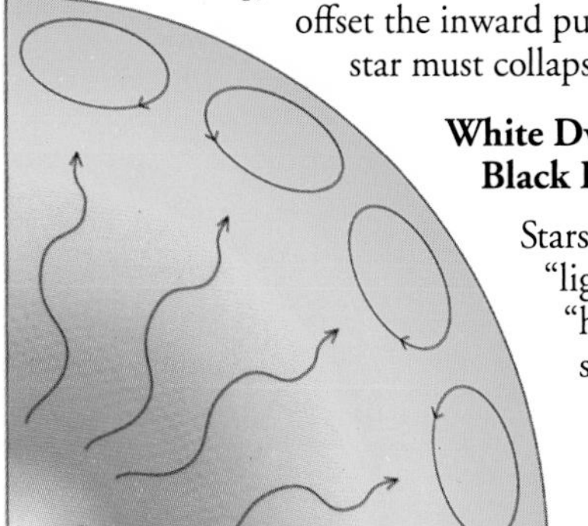

White Dwarfs and Black Dwarfs

Stars can be "lightweights" or "heavyweights." The smallest are only a tenth as massive as our sun; the biggest can weigh one thousand times

The first white dwarf to be discovered, Sirius B was so named because it is the unseen companion of Sirius A, the brightest star in the night sky—so brilliant, in fact, that its glare long made Sirius B impossible to see. Astronomers finally photographed Sirius B in 1863 after repeated attempts. Although ten thousand times fainter than its companion, it was, they learned, just as hot (10,000 degrees K). A star with such characteristics would have to be extremely small, comparable to earth in size. The term "white dwarf" was coined. In the 1930s American astrophysicist Subrahmanyan Chandrasekhar discovered that white dwarfs are the remnants of stars of less than 1.4 solar masses that run out of nuclear fuel and start to collapse on themselves. Such stars are neither hot enough nor dense enough at the core to trigger thermonuclear fusion, or "burning," of carbon or oxygen. Iron—an element found in meteorites (thin section, left)—can be forged only in the cores of much more massive stars.

more. The death of a star can be a quiet, drawn-out affair or a sudden, violent event. It all depends on the star's initial mass.

What Lies in Store for Our Sun? Will Our Descendants Have to Find Another?

Nine billion years from now, when it has run through its reserve of nuclear fuel, the sun will shrink under its own gravity to about the size of the earth. It will have become a dwarf star about 7500 miles in diameter. Its temperature will rise as the energy of contraction is converted into heat. When it becomes white-hot, it will officially be a "white dwarf."

The matter packed inside such a star is superdense: A teaspoonful weighs a ton. As its core shrinks, the star's outer layers will drift away into space. Set aglow by the central white dwarf, the surrounding gaseous material will look like a ring with splashes of red, green, and yellow. (This is called a "planetary nebula," a misleading term because the phenomenon it describes has nothing whatever to do with planets.)

Our descendants will have lost their source of energy. The time will have come for them to set out in search of another sun. Perhaps the age of galactic

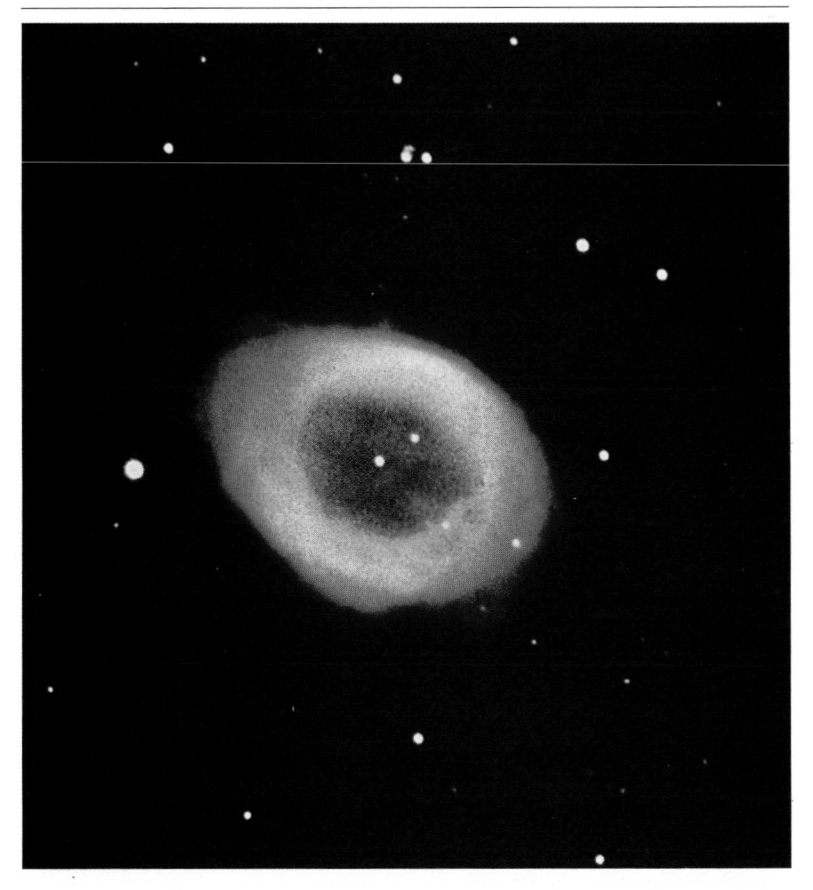

colonization we read so much about in science fiction will dawn at last.

It takes white dwarfs billions of years to radiate away their heat. Eventually the white dwarf will fade into an invisible "black dwarf" and take its place among the countless burned-out stellar corpses that litter the galactic immensities. By then its planetary nebula will have dissipated, scattering into space the heavy elements forged in once-fiery stellar crucibles. This protracted death awaits all stars with a mass no more than 1.4 times the sun's.

The Ring Nebula in Lyra (M57), 4000 light-years from earth, glows with radiation emitted by a white dwarf, the bright point of light at its center. This planetary nebula will no longer be visible some 50,000 years from now.

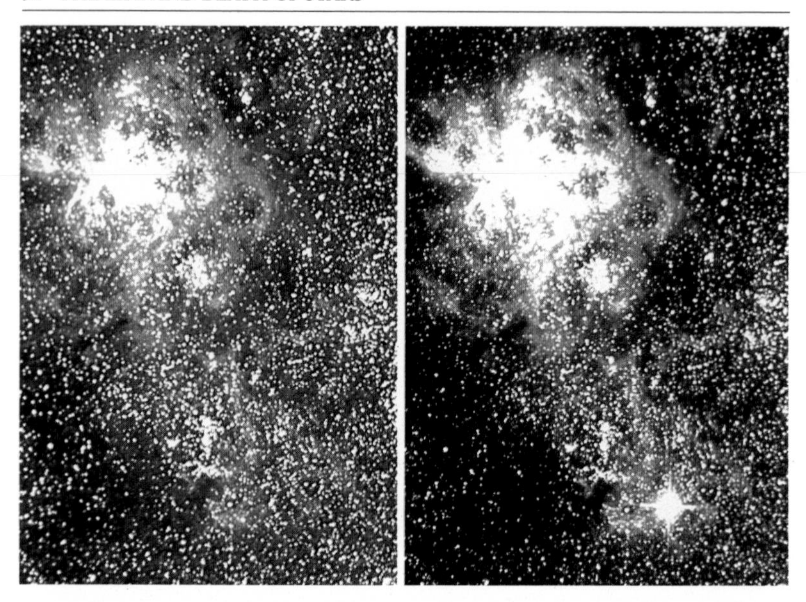

The Explosive Death Throes of Supernovas

What happens to more massive stars? Their death struggle is far more violent; but the final outcome, once again, hinges on their initial mass—specifically, whether or not its quantity exceeds about five solar masses.

Let's first chronicle the death of a star that is from 1.4 to about five times more massive than our sun. The large mass of the star causes it to shrink down to a radius of about 6 miles. All of its matter is converted into neutrons: A teaspoonful of neutron star material weighs a billion tons. You could obtain the same mind-boggling density by squeezing a hundred Eiffel Towers into a space the size of the ball at the tip of your pen.

The catastrophic implosion of the core triggers a dazzling explosion. The outer layers are blown away into space at thousands of miles per second. The explosion shines as brightly as billions of suns. A point of light almost as luminous as an entire galaxy flares in the sky. This is called a "supernova." These explosive deaths occur about once every century in any given galaxy, and every second if we take into account the hundred billion

The appearance a few years ago of a new star in the Large Magellanic Cloud signaled the explosive death of a massive star (in the lower righthand corner of the photograph at right). Christened supernova 1987A, it was so bright that observers in the southern hemisphere could see it with the naked eye for 6 months. It has since faded, but astronomers continue to study its death throes with large telescopes. The explosion occurred 150 thousand years ago, in pre-Neanderthal times, but we have only just now learned of it.

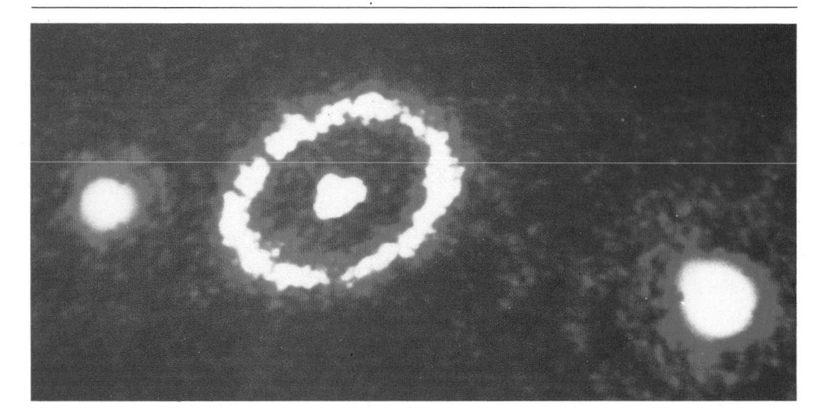

galaxies in the universe. Since the dawn of recorded history, humans have witnessed about ten supernovas in the Milky Way. The "new star" young Tycho Brahe observed in 1572 in the constellation Cassiopeia—the one that prompted him to doubt the immutability of the heavens—was just such an event. The remnant of that supernova now bears his name.

On 23 February 1987, a supernova appeared in the Large Magellanic Cloud, a dwarf galaxy 150,000 light-years away in orbit around the Milky Way. It sent shock waves through the world of astronomy. Scientists seized the opportunity, marshaled the full panoply of state-of-the-art equipment—large-aperture ground-based telescopes, space satellites, neutrino detectors—and, for the first time ever, were able to scrutinize the death of a nearby star with precision.

A Hubble Space Telescope image reveals a ring of matter 1.3 light-years across surrounding SN 1987A (bright dot at the center of the ring). Glowing with the supernova's intense radiation, the ring is thought to consist of material the progenitor star shed during its red giant phase, 10,000 years before the explosion. Before that, it was a blue supergiant ten times bigger, twenty times more massive, and 100,000 times brighter than our sun.

The Crab Nebula: A Guest Star

One of the most famous supernovas in the annals of astronomy was the explosion responsible for the star remnant we now call the Crab Nebula. On the morning of 4 July 1054, an object flared in the sky; Chinese astronomers referred to it poetically as a "guest star." It shone so brightly that weeks later it was still visible in daylight. Yet there is no mention of the event in contemporary European chronicles, perhaps because people in the West were "blinded" by their belief that the stars were changeless. The "guest star" is no longer visible to the unaided eye, although with a telescope an observer can make out the faint remnant with the crablike shape that inspired its name.

But the Crab Nebula was destined to bask in even greater celebrity. In 1967 astronomers discovered that its central region harbors a neutron star, which they detected as bursts of energy flashing on and off thirty times a second. Two factors, they learned, conspire to produce this extraordinary phenomenon, commonly known as a pulsating star, or "pulsar," for short. First, neutron stars do not emit radiation over their entire surface,

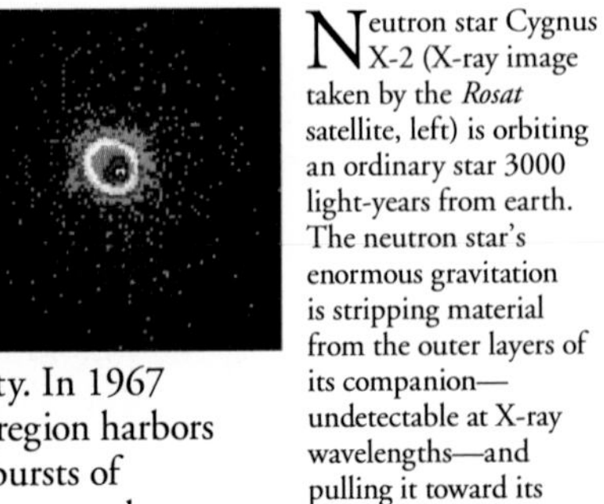

Neutron star Cygnus X-2 (X-ray image taken by the *Rosat* satellite, left) is orbiting an ordinary star 3000 light-years from earth. The neutron star's enormous gravitation is stripping material from the outer layers of its companion—undetectable at X-ray wavelengths—and pulling it toward its surface. As the stellar material heats up, it emits X rays in huge quantities.

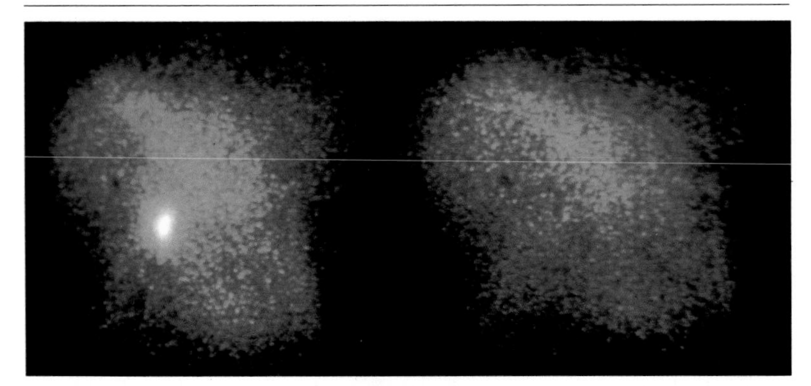

but in twin beams. Second, they rotate very rapidly. Just as an ice skater can spin more quickly by drawing the arms in toward the body, collapsed stars rotate much more rapidly than do stars of ordinary size. Every time one of a pulsar's beams sweeps across the earth, it is observed as a flash of light as if from some great celestial lighthouse.

Like a spinning top, the rotating Crab Nebula pulsar will gradually decelerate as it uses up the energy stored up during its gravitational collapse. Several million years from now it will finally flicker out, a stellar corpse never to be seen or heard from again.

A neutron star (X-ray image, above) is flashing on and off at the core of the Crab Nebula (optical photograph, opposite below), a supernova remnant. Products of stellar alchemy within the nebula's red and yellow filaments are spreading outward through space at thousands of miles per second.

Black Holes: The Ultimate Star Death

What happens to a star of more than five solar masses? Once it has exhausted its fuel, it will be crushed into an extremely compact state, creating a gravitational field so powerful that it bends space over on itself and traps light forever. The star will become a black hole. If light cannot escape it, neither can anything else. All matter caught

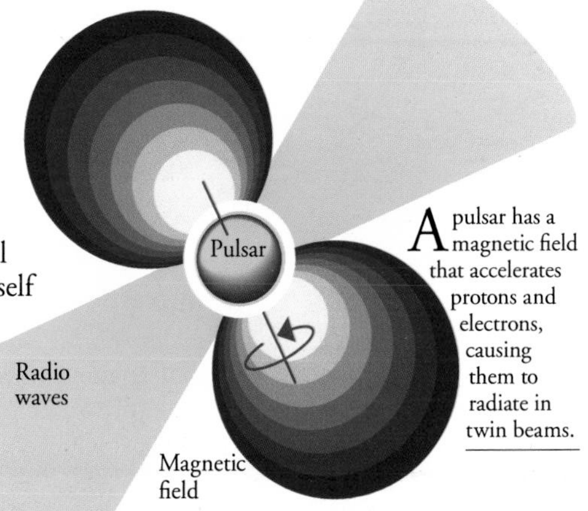

A pulsar has a magnetic field that accelerates protons and electrons, causing them to radiate in twin beams.

in a black hole's grip is doomed to a one-way trip.

Theoretically, any object can turn into a black hole. It simply has to be compressed to a point beyond which light will be turned back by the pull of gravitation. You, too, would become a black hole if a pair of gigantic hands could squeeze you to less than 10^{-23} of an inch, a speck a million billion times smaller than an atom. If the earth were to shrink to the size of a billiard ball, it, too, would become a black hole. But black holes are rare because compression of that magnitude is difficult to achieve. The electromagnetic force, which binds atoms and molecules together and marshals them into crystal lattices, fiercely resists such extreme concentration. Creating a black hole requires the gravity of a star that is more than five times as massive as our sun.

If a black hole prevents light from escaping, how can astronomers deprived of that precious communications link detect its presence? Once it has formed, a black hole pulls in

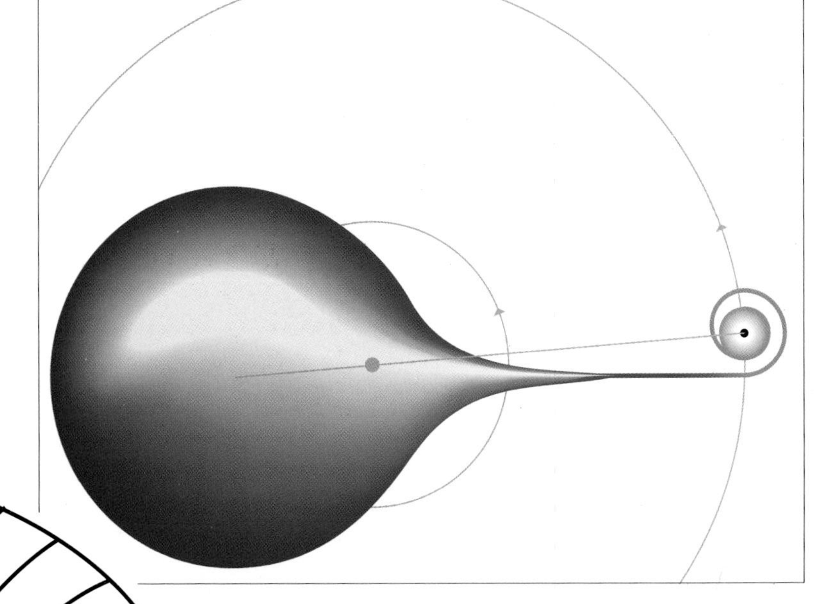

and consumes everything that strays within range, getting bigger and more massive in the process. That frenzied cannibalism is precisely what gives it away.

Many massive stars are in binary systems in which one star is in orbit around another. Should the more massive of the two stars collapse and form a black hole, the other will continue to circle its unseen companion as if nothing has happened. The gravitational field controlling the motion of the visible star depends solely on the pair's total mass, which remains constant. But the black hole's intense gravity field must wreak some havoc. It draws the gaseous atmosphere of the visible star toward it. The gas particles hurtle toward the black hole at enormous speed. As they smash into one another, their atoms release tremendous amounts of X rays. The accreted material then vanishes forever into the belly of the beast.

Space that is a safe distance from a black hole is "flat," that is, not curved. The gravitational field created by a black hole is so strong that it warps space around itself (diagram, opposite). Anything within range is sent hurtling into the abyss to certain destruction. In the constellation Cygnus, an optically invisible companion emitting tremendous quantities of X rays is orbiting a blue supergiant (diagram, above). Astronomers suspect that the unseen object in this system is a black hole.

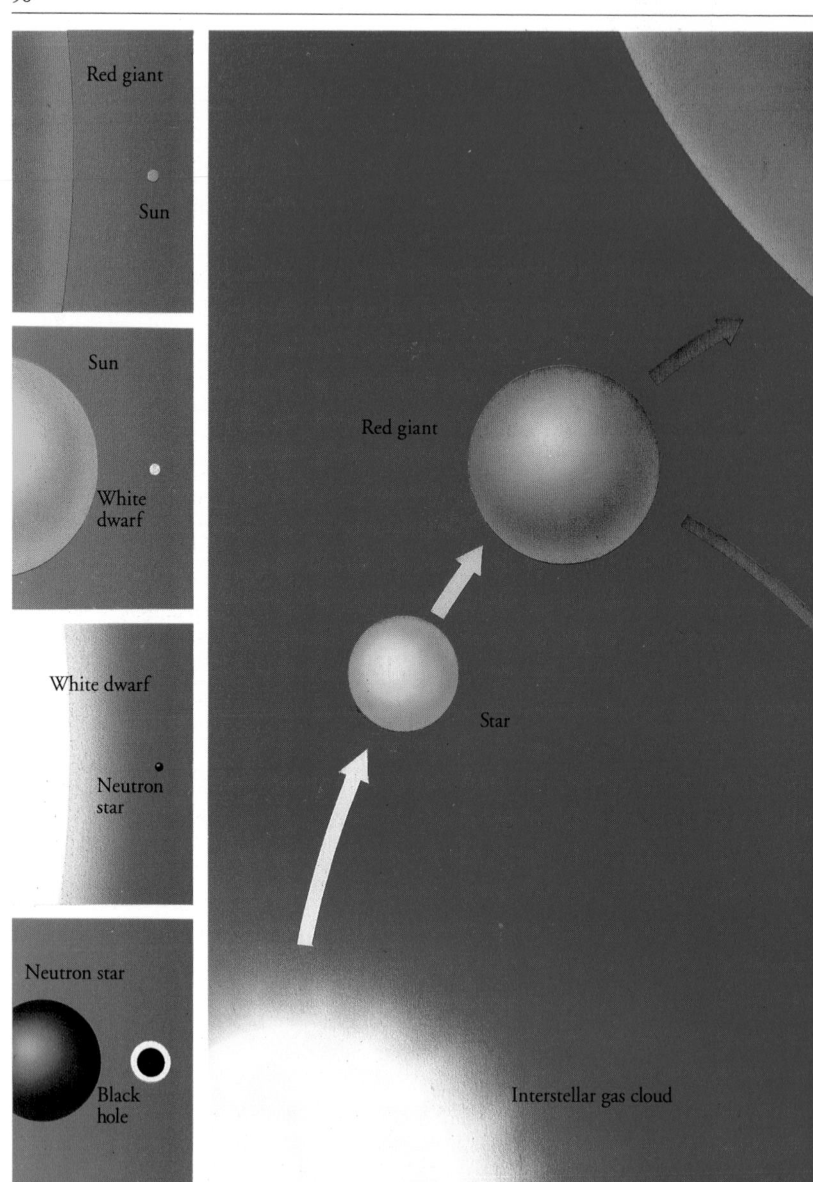
Red giant
Sun
Sun
White
dwarf
White dwarf
Neutron
star
Neutron star
Black
hole
Red giant
Star
Interstellar gas cloud

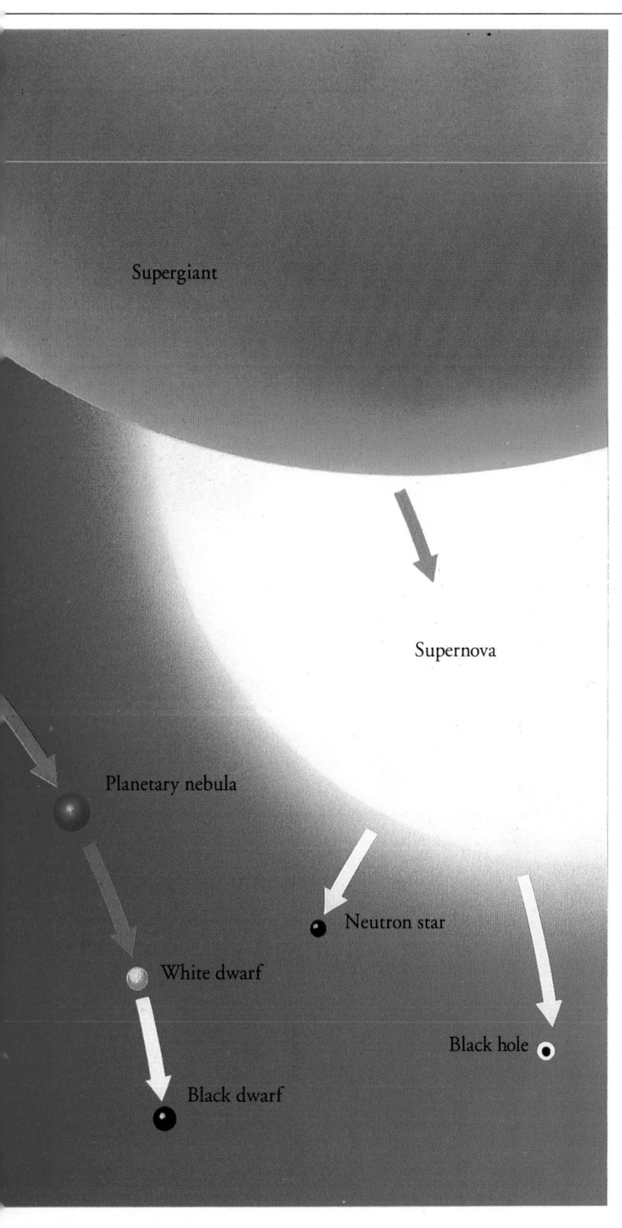

A star's ultimate fate depends on how well it can hold out against the effects of gravity. As long as a star has fuel to burn, the outward radiation pressure produced by nuclear fusion in its core will support the star against the compressive force of its own gravity. Once it has exhausted its fuel supply, however, gravity takes over and the star begins to contract. The electrons in a star of less than 1.4 solar masses resist being too densely packed and exert sufficient pressure to halt compression at a radius of about 3700 miles. Thus, electrons support white dwarfs against their own gravity. In stars of 1.4 to five solar masses, electrons cannot resist stellar contraction, leaving the outward pressure of tightly packed neutrons as the next line of defense against the crush of gravity. They stabilize neutron stars at a diameter of roughly 12 miles. If the mass of the star exceeds five times that of the sun, neither electrons nor neutrons can stave off further gravitational collapse. The star will continue to shrink and become a black hole.

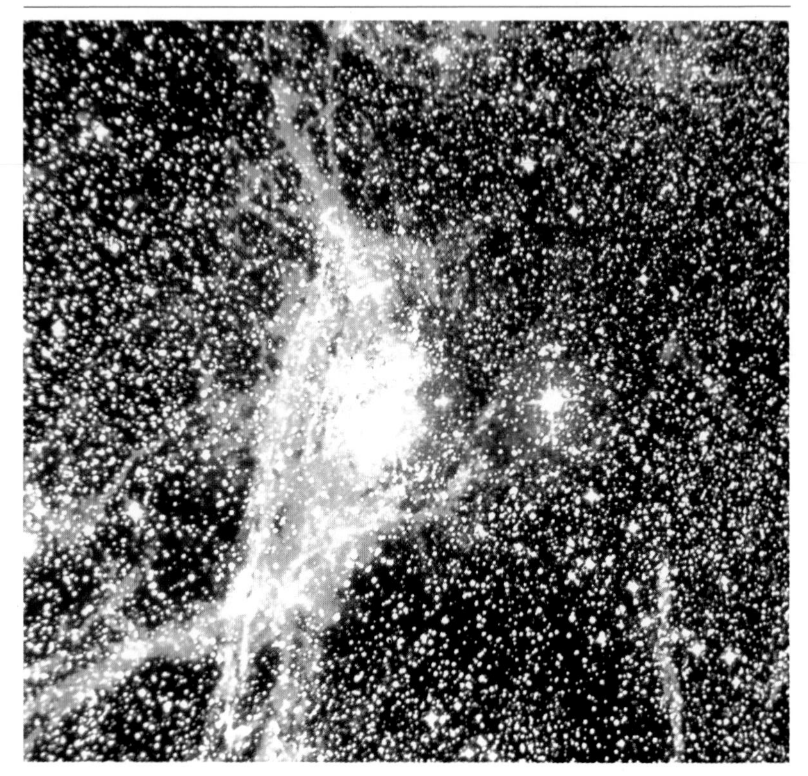

Debris left over from a stellar explosion 12,000 years ago can still be seen in the constellation Vela (above). Such supernova remnants are scattered throughout the Milky Way.

How Supernovas Did Us a World of Good

Like neutron stars, black holes are produced amid the tremendous explosion of supernovas. Stars have been able to manufacture elements heavier than helium, which was fine as far as it went. But what would have been the point of all that wonderful nuclear alchemy if its products had remained captive inside stellar crucibles? To reach the next level of complexity—synthesizing complete atoms—the electromagnetic force had to find a way to bind electrons in orbit around nuclei. This process could not take place in the interiors of stars. Their tremendous heat flung atoms into each other, shattering them the instant they formed. Atom breeding required cooler, calmer surroundings. Ranging in temperature from a frigid 50 degrees K to a scorching 10,000 degrees K, the interstellar medium was the ideal location.

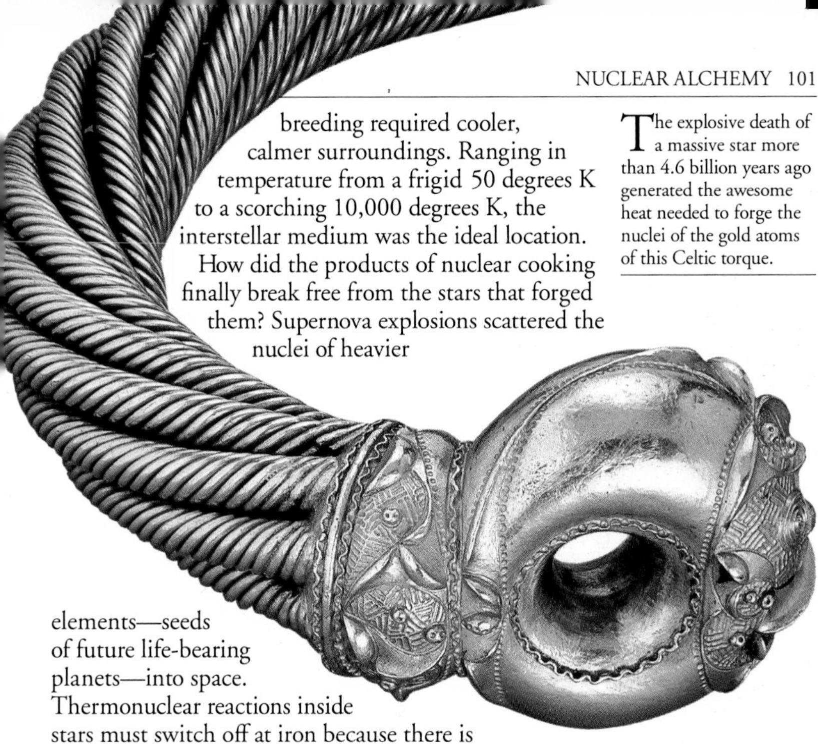

How did the products of nuclear cooking finally break free from the stars that forged them? Supernova explosions scattered the nuclei of heavier elements—seeds of future life-bearing planets—into space. Thermonuclear reactions inside stars must switch off at iron because there is not enough energy to sustain them. Supernovas, however, had energy to spare. Superheating in the exploding star's outer envelope set off a chain of fusion reactions and synthesized all the elements heavier than iron, including silver, gold, lead, and uranium.

Supernovas also threw high-energy electrons and protons into interstellar space. Some of those fast-moving "cosmic rays" would one day reach the earth and alter genetic structure. Therefore, supernovas can also be credited with the mutations that made it possible for life on earth to evolve from rudimentary cells.

The explosive death of a massive star more than 4.6 billion years ago generated the awesome heat needed to forge the nuclei of the gold atoms of this Celtic torque.

However, the iron atoms of this 1st-century Roman chisel (below) were synthesized in the multibillion-degree heat of a stellar core while the star was still "alive and well." Then, during its explosive death, they were released fully formed into interstellar space and sent surging toward their ultimate destination, the earth's crust.

On the blue planet, nature played the card of life. Of the countless plant and animal species that evolved on earth, the very one capable of marveling at the beauty and harmony of the universe—humankind—is jeopardizing the ecological balance of the biosphere. We had better take good care of our planet. In the vastness of the cosmos, another one fit to live on will be hard to come by.

CHAPTER V
A PLANET IS BORN

Early in its history, the earth was covered with fiery lava like the kind flowing from Hawaii's Kilauea volcano (opposite). Right: The ancestors of this primitive organism, a siphonophore, evolved about 500 million years ago, more than 3 billion years after life first appeared on earth.

Molecules in Space

Creative alchemy inside massive stars supplied the universe with heavy nuclei, which the explosive force of supernovas then scattered into interstellar space. But before they could be processed into atoms and molecules there had to be some way of helping nuclei to collide and bond. In general, interstellar clouds were too rarified to serve as atomic breeding sites. But interstellar dust grains, condensed out of the atmospheres of red giants and expelled into space by intense radiation, saved the day.

These solid particles, a tiny fraction of an inch across, were giants on the atomic scale: billions of oxygen, silicon, magnesium, and iron atoms marshaled by the electromagnetic force into a rigid network, forming a solid core with a thin coating of ice. The surface of interstellar dust grains proved fertile breeding sites indeed. Held together, once again, by the electromagnetic force, the nuclei of heavy elements forged deep within stars repeatedly combined.

Molecules containing two, three, four, and up to twelve atoms assembled in interstellar space. Especially abundant were hydrogen, carbon monoxide, and the water molecules that were to prove so crucial to the development of life on our blue planet. Then came the methane and ammonia molecules later responsible for the noxious effluvia in earth's primeval atmosphere.

Radio astronomers have detected nearly a hundred types of molecules in the interstellar medium; hydrogen, carbon, nitrogen, and oxygen seem to predominate. These four elements make up more than 99 percent of all living things. Needless to say, the complexity of the DNA double helix was still a long way off.

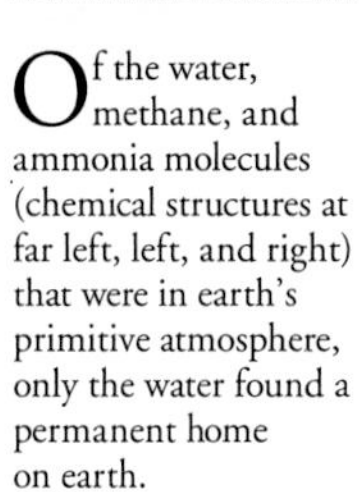

Of the water, methane, and ammonia molecules (chemical structures at far left, left, and right) that were in earth's primitive atmosphere, only the water found a permanent home on earth.

Interstellar dust grains (chemical structure, above) were the first solid bodies to form in the universe.

The Birth of the Solar System

The next step was to build atomic structures involving not tens or thousands of atoms, but millions of them. But how? That process required far more hospitable breeding grounds than the interstellar medium. Nature then contrived planets.

Let's take a close look at how our beloved home planet came into being. The cosmic clock struck 10.4 billion years. All that time the universe had been expanding, thinning out, and cooling down. Galaxy groups, clusters, and superclusters had been woven together into a cosmic tapestry. Several generations of stars had lived and died, dispersing heavy elements into the interstellar medium.

The Universe Now Harbored Hundreds of Billions of Galaxies

About a third of the way in from the edge of one such galaxy, the Milky Way, an interstellar gas cloud began to contract. Its interior temperature soared to 10 million degrees K, setting off the nuclear fusion of hydrogen. The gas cloud ignited and became a star. Our sun, a third-generation star, had switched on.

During the contraction phase, dust grains condensed out of the gas cloud and began to whirl about the sun; eventually they settled into rings like the ones encircling

Saturn. Within these rings, dust grains with slightly greater mass—hence, greater gravitational pull—started to coalesce into larger clumps. Their mass and gravity increased, and the pace of accretion quickened. In time, gravitational forces aggregated nearly all of the material in the rings into the nine spherical bodies we call planets. Smaller clumps condensed to form satellites around every planet except Mercury and Venus. The solar system had been born.

Our Planet Earth

Any leftover dust coalesced into the asteroids, thousands of jagged lumps of rock not massive enough to aggregate gravitationally into spheres. Ranging from less than an inch to several miles in diameter, these "minor planets" now circle the sun in the asteroid belt between the orbits of Mars and Jupiter. When the solar system was still very young, a great many asteroids crashed into the freshly wrought planets. Extensive

The solar system (formation sequence, opposite). The sun's closest neighbors, Mercury, Venus, Earth, and Mars, are relatively small and have rocky surfaces rich in heavy elements and thin or nonexistent atmospheres. Farther out lie the massive Jupiter (above), Saturn, Uranus, and Neptune, which have no solid surface. Their thick atmospheres consist primarily of light elements, such as helium and hydrogen. Pluto belongs to neither group. Its origin is still unknown. Overleaf: The solar system.

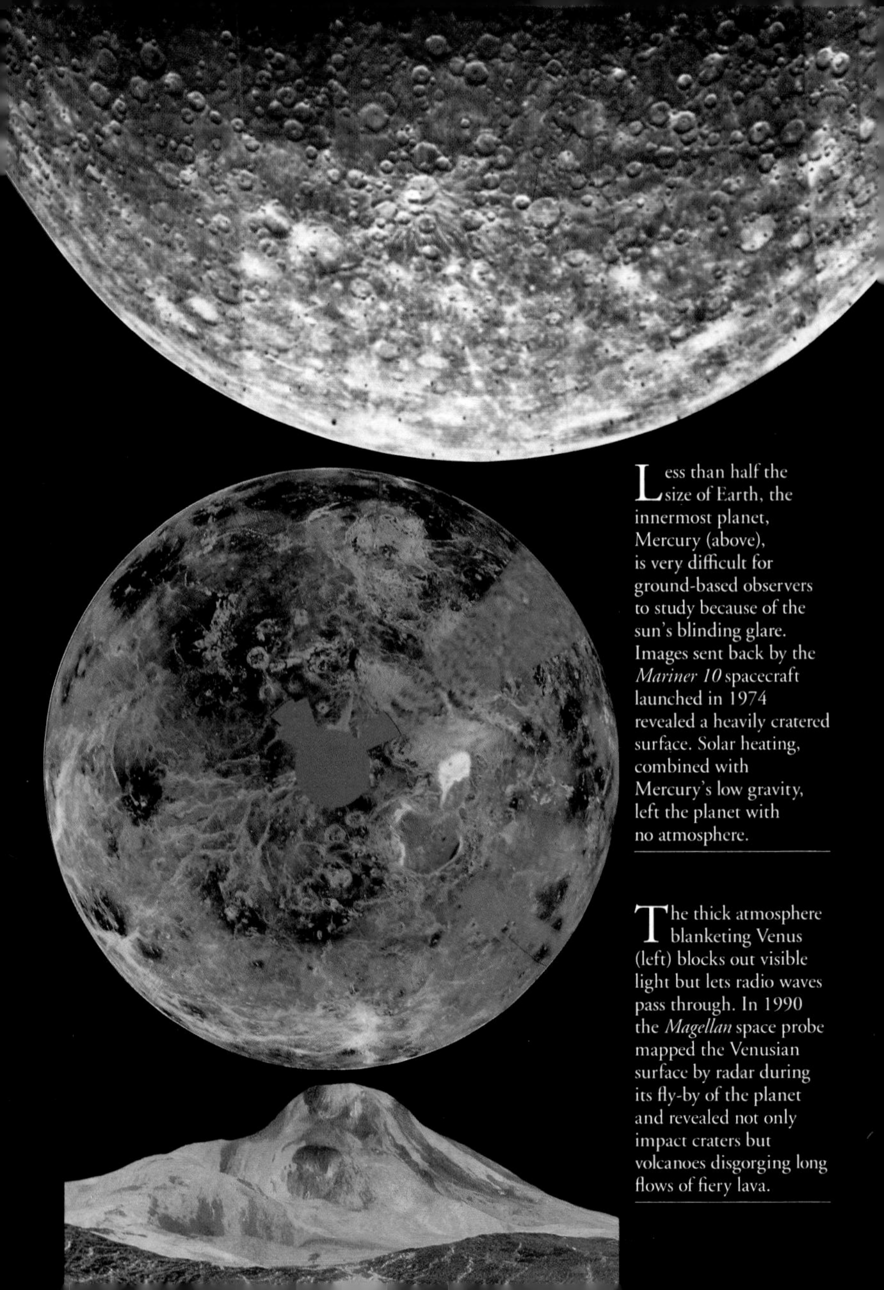

Less than half the size of Earth, the innermost planet, Mercury (above), is very difficult for ground-based observers to study because of the sun's blinding glare. Images sent back by the *Mariner 10* spacecraft launched in 1974 revealed a heavily cratered surface. Solar heating, combined with Mercury's low gravity, left the planet with no atmosphere.

The thick atmosphere blanketing Venus (left) blocks out visible light but lets radio waves pass through. In 1990 the *Magellan* space probe mapped the Venusian surface by radar during its fly-by of the planet and revealed not only impact craters but volcanoes disgorging long flows of fiery lava.

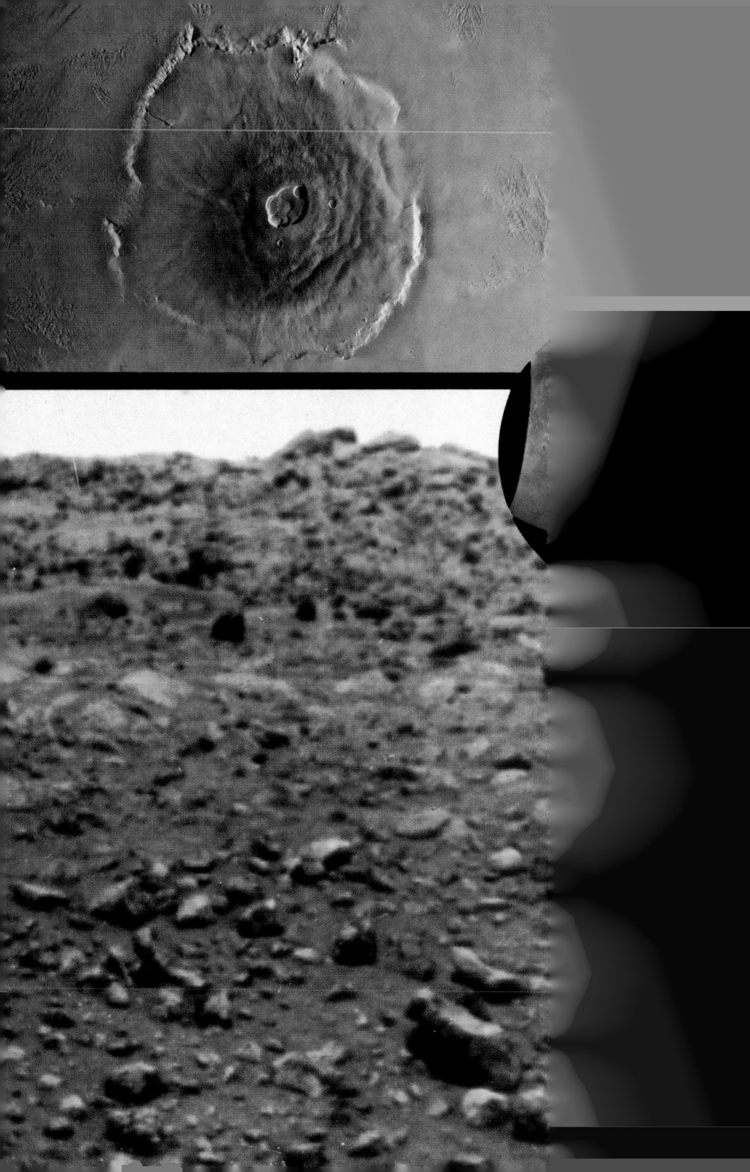

Jupiter is the colossus of our solar system. It is 318 times more massive than Earth and two and a half times more massiv than all the other planet and satellites combined. Jupiter has the shortest rotational period of any planet even though it measures eleven Earth diameters across. Its rocky core is shrouded ir a mostly hydrogen and helium cloud layer more than 12,000 miles thick. Because of Jupiter's rapi axial rotation, violent winds continually churn the atmosphere, generating zones of high pressure (light areas) anc belts of low pressure (dar areas) parallel to its equator (left below). The Great Red Spot (left above), a swirling disturbance so huge that three Earths could fit inside it, is the planet's most conspicuous surfac marking. Images transmitted by the *Voyager* probes revealed the astonishing variety o Jupiter's four Galilean satellites (above, top to bottom: Mercury-sized Ganymede and Callisto, Moon-sized Io and Europa). Impact craters pock the surface of Ganymede and Callisto; faults scar frozen Europa scorching Io is convulsec by volcanic eruptions.

Saturn's most celebrated feature, its ring system (above), is actually made up of countless rock and ice fragments ranging in size from snowflake-size specks to boulders tens of yards across. Each and every piece is in an independent orbit in the planet's equatorial plane (left). Scientists speculate that gravitational forces prevented bits of debris too close to the still-forming central planet from coalescing into satellites.

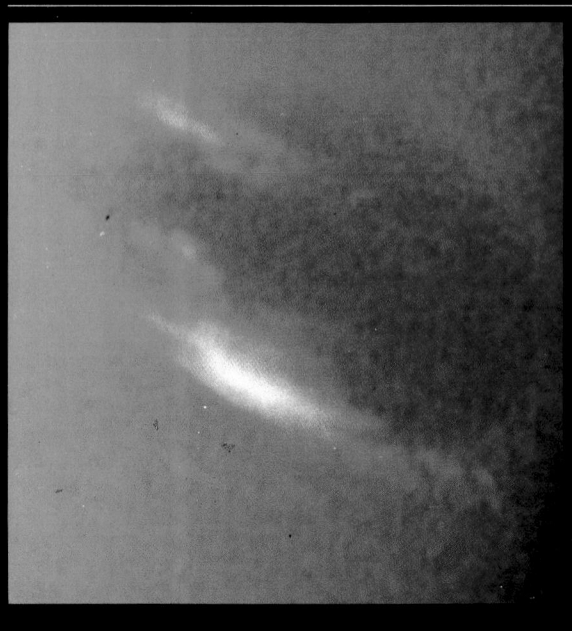

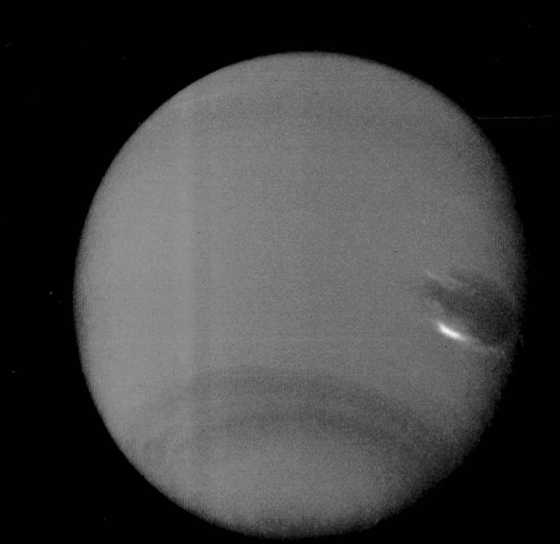

Nineteenth-century English mathematician John Couch Adams and astronomer Urbain Leverrier of France independently predicted the existence of Neptune (left below). The only way they could account for perturbations in the orbit of Uranus was to postulate the gravitational pull of an unseen eighth planet. They calculated Neptune's position using Newton's laws of gravitation, and in 1846 the planet was discovered exactly where it had been predicted. The *Voyager 2* space probe flew by the planet in August 1989. Neptune is shrouded in a methane-rich cloud layer that absorbs yellow and red light and reflects only blue wavelengths. Neptune's atmosphere, like Jupiter's, is buffeted by violent storm systems like the nearly Earth-size Great Dark Spot (detail, above). The frigid temperature of the Neptunian atmosphere (–213 degrees C) freezes methane gas into crystals that condense as whitish, cirrus-like clouds.

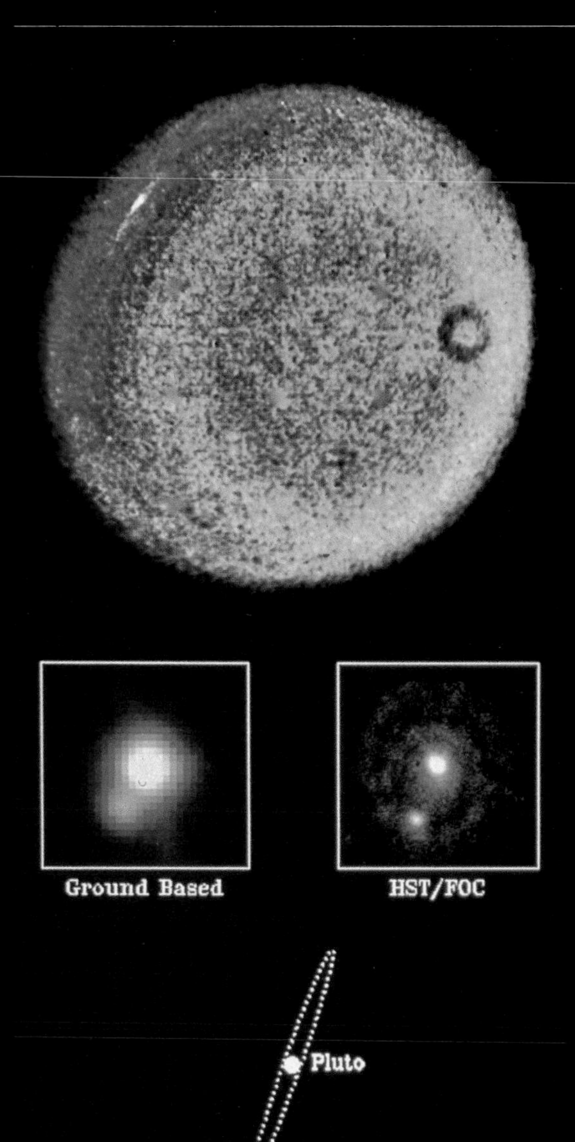

Uranus (left above) is 1.8 billion miles from the sun. Its rotational axis is so strongly tilted that its poles lie nearly in its orbital plane. The resulting seasons are extreme to say the least: Of the 84 Earth years it takes the planet to complete one revolution around the sun, 42 are spent in summer light and 42 in winter darkness. Some believe that a close passage by a massive asteroid may have been responsible for tipping Uranus. The ninth planet in the solar system, Pluto, was not discovered until 1930. Its highly eccentric orbit, inclined at 17 degrees to the plane of the solar system, is unlike that of any other planet. This unique feature has prompted speculation that Pluto is a former Neptunian satellite sent into an independent orbit around the sun after interacting with a massive object. A scant 12,200 miles separate Pluto from its moon, Charon, which was discovered in 1978. An Earth-based telescope imaged them as an elongated blob of light (left inset), but the Hubble Space Telescope had no trouble resolving them into separate bodies (right inset).

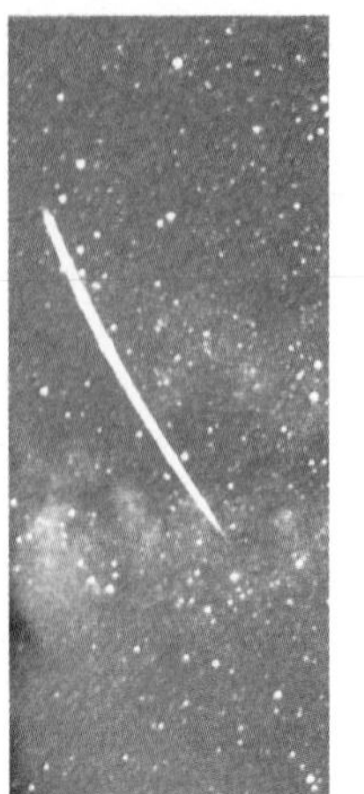

cratering on the Moon and Mercury bear mute witness to this activity. On Earth, the erosive effects of rain, rivers, glaciers, and continental motion have erased all traces of this period of heavy bombardment.

A Number of Recent Collisions Have Left Scars

One example is the stupendous Barringer meteor crater in Arizona, a huge depression more than half a mile across which an impact blasted out of the earth's crust some 30,000 years ago.

Bits of cosmic debris enter the earth's atmosphere from time to time. As atmospheric friction burns them up, meteors leave fiery trails in their wake: These are the spectacular shooting stars we see streaking across the starry night sky. By the time they land on the ground, they are reduced to charred lumps of rocky material. Some specimens are sent to laboratories for analysis, because whatever science can learn from

Every day, 300 tons of cosmic debris, burning up as it enters the earth's atmosphere, rain down on the earth (photograph, left). Most bits of matter that actually land on the ground are too small and light to do serious damage. One exception was the object that gouged out the Barringer meteor crater in Arizona (below). Measuring about 160 feet across, it crashed into the earth at a speed of 25,000 miles per hour and released energy equivalent to the explosion of a 20-megaton hydrogen bomb.

their chemical composition helps shed light on the processes that shaped the solar system.

A Tale of Water and Deluge

Now that nature had formed planets, the stage was set for the emergence of life. This task called for the cooperation of a powerful ally, water, which had already been assembled in the frigid depths of interstellar space from bits of expelled stellar material.

A billion years had elapsed since the birth of the sun. Volcanoes had already disgorged red-hot lava flows over the surface of our planet, and the earth had cooled substantially. As it solidified and a protocontinent took shape, the lava gave off huge amounts of gas, blanketing the earth with an atmosphere a hundred times thicker than today's.

The early atmosphere of hydrogen, ammonia, methane, water vapor, and carbon dioxide was not suitable for life. As the planet continued to cool, water condensed out of the atmosphere, and torrential rains inundated the planet, leaving three-quarters of its surface covered with oceans.

The Secret of Immortality

Water now began to fulfill its mission as the catalyst of life. Because water is a highly effective dissolving agent, all sorts of alien molecules could readily adhere

The red-hot lava disgorged by volcanoes (left) is molten rock welling up from deep inside the earth. Early in its history, our planet was covered with lava. Two factors conspired to produce the tremendous heat associated with volcanism: a period of intense bombardment by asteroids when the solar system was still forming, and the thermal energy that is released as radioactive elements spontaneously decay in the earth's interior. This heat is responsible for continental drift, which pushes tectonic plates against one another and thrusts up entire mountain ranges. The Himalayas, for example, were uplifted when the plates carrying India and Eurasia collided. Compressional stress along plate boundaries can build to intense levels, with literally earth-shaking consequences. One example, California's notorious San Andreas Fault, is located where the North American and Pacific plates are sliding past each other.

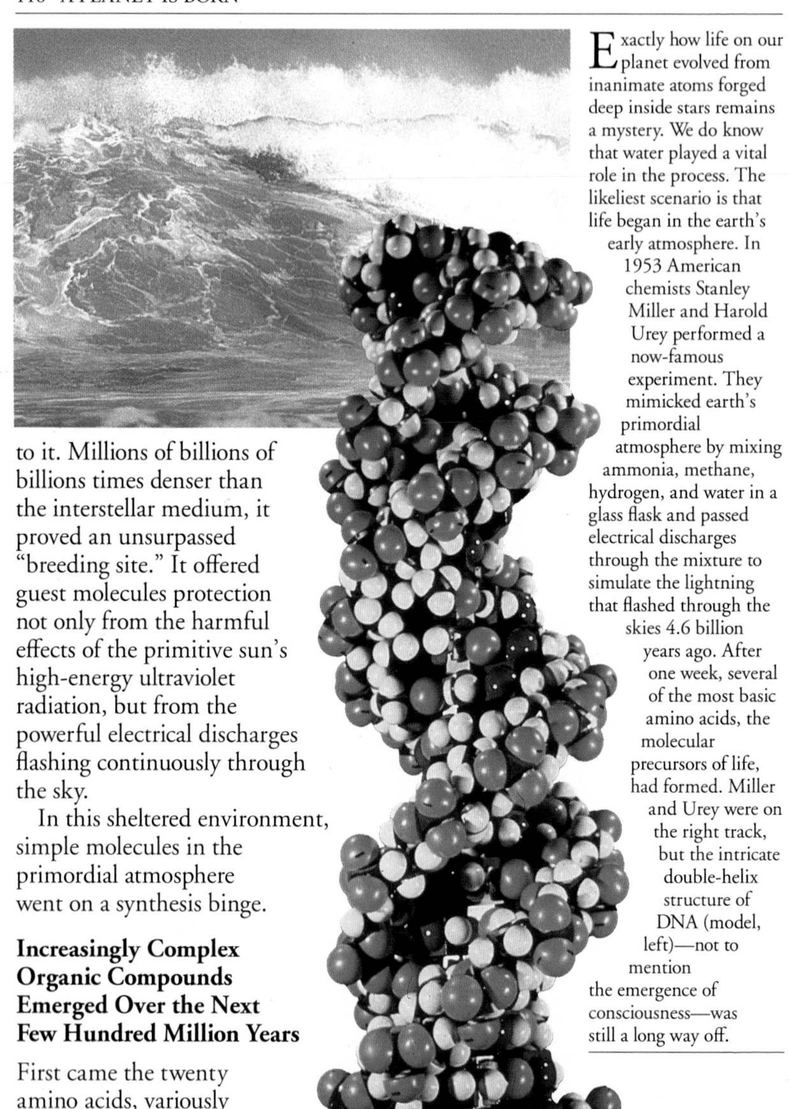

to it. Millions of billions of billions times denser than the interstellar medium, it proved an unsurpassed "breeding site." It offered guest molecules protection not only from the harmful effects of the primitive sun's high-energy ultraviolet radiation, but from the powerful electrical discharges flashing continuously through the sky.

In this sheltered environment, simple molecules in the primordial atmosphere went on a synthesis binge.

Increasingly Complex Organic Compounds Emerged Over the Next Few Hundred Million Years

First came the twenty amino acids, variously arranged clusters of about thirty atoms each. Amino acids then assembled into

Exactly how life on our planet evolved from inanimate atoms forged deep inside stars remains a mystery. We do know that water played a vital role in the process. The likeliest scenario is that life began in the earth's early atmosphere. In 1953 American chemists Stanley Miller and Harold Urey performed a now-famous experiment. They mimicked earth's primordial atmosphere by mixing ammonia, methane, hydrogen, and water in a glass flask and passed electrical discharges through the mixture to simulate the lightning that flashed through the skies 4.6 billion years ago. After one week, several of the most basic amino acids, the molecular precursors of life, had formed. Miller and Urey were on the right track, but the intricate double-helix structure of DNA (model, left)—not to mention the emergence of consciousness—was still a long way off.

long chains to make proteins, which in turn linked up to form intertwined double chains of DNA molecules.

Millions of atoms long, these strands held the secret of immortality: self-replication. They were destined to carry the genetic code of all living things. When the cosmic clock struck 11.5 billion years, naked chains of DNA evolved into cells containing millions of billions of atoms each. The primeval ocean teemed with single-celled bacteria and algae.

The Ascent Toward Life

There was a pause of 3 billion years. Then, when the cosmic clock struck 14.4 billion years—600 million years ago—nature indulged its passion for organizing

This inspiring image of earth floating in the blackness of space is a reminder of how unique and fragile our planet is—unique because the oceans that give our planet its bluish color cannot be found anywhere else in the solar system and fragile because recently we humans have begun to tamper with its ecological balance. The results have been dramatic, possibly irreversible.

as never before. Jellyfish, mollusks, crustaceans, and fish all evolved in just a few hundred million years.

One Hundred and Fifty Million Years Later, Flora Carpeted the Earth

Through photosynthesis, green plants tapped the energy of sunlight and converted their constituent elements into carbohydrates, in the process releasing oxygen into the air. Triplets of atmospheric oxygen atoms combined to form ozone. The resulting ozone layer absorbed the sun's harmful ultraviolet radiation, allowing organisms to emerge from the protection of the water and live on dry land.

The first reptiles may have appeared 300 million years ago, with dinosaurs and birds coming on the scene about 75 million years later. When the dinosaurs went extinct, after ruling the earth for about 165 million years, the way was cleared for the proliferation of mammals. The primates evolved some 20 million years ago, and about 2 million years ago the first representatives of *Homo sapiens* trod the earth.

It took 15 billion years to progress from high-energy nothingness to human beings made up of 30 billion billion billion particles and endowed with brains built up from billions of neurons.

According to 19th-century English naturalist Charles Darwin, the evolution of living species from primitive cells to human beings came about through random genetic mutation and the mechanism of natural selection. Those species better adapted to their environment survived and passed on favorable variations to successive generations; others died out. To defend themselves against predators, some early organisms, such as crustaceans and mollusks, developed hard shells.

Ice Boxes and Furnaces

Is life confined to earth? Are humans the only intelligent beings in the universe? But for them, would its secret melody go unheard? To all appearances, not one of the other eight planets in our solar system can sustain life. The *Viking* space probes explored the likeliest candidate, Mars, but found no living organisms, much less intelligent beings. The other planets look equally unpromising; they are too scorching or too frigid, their atmospheres too stifling or too rarified. Life is a fragile thing that requires careful treatment: Without mild, benign surroundings, it cannot develop.

Since terrestrial life is the only kind we have to go by, all speculation about extraterrestrial life is inescapably

The emergence of better-adapted species did not necessarily doom those preceding them to extinction. They coexisted. These artist's conceptions of Early Jurassic life (about 180 million years ago) show ammonites in the water, dragonflies in the air, and reptiles on land (below and opposite above).

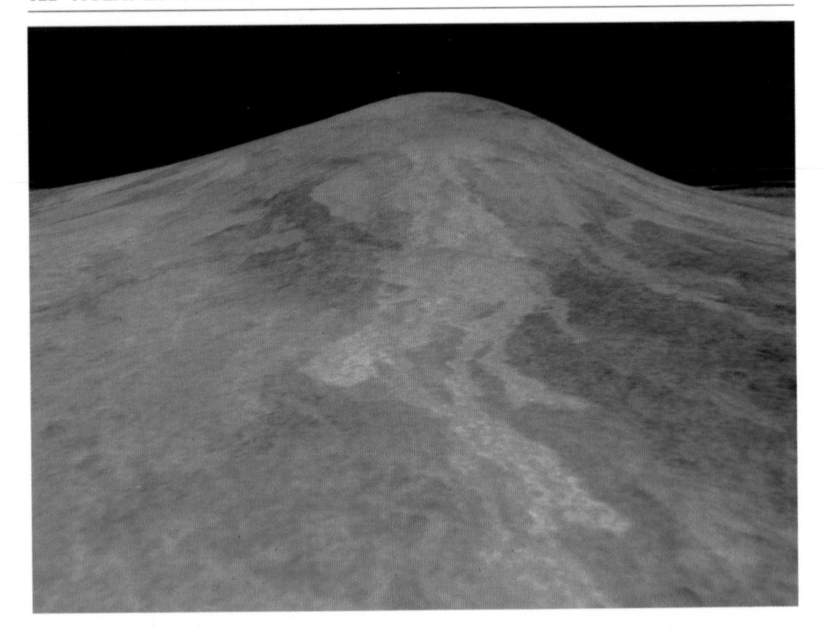

anthropomorphic. An ideal life-sustaining environment would possess water and a surface temperature of between 0 and 100 degrees C. For these conditions to prevail, a planet has to be very carefully positioned with respect to the sun. Had Earth formed farther afield than Jupiter, it would have turned into a frigid realm of ice and frost not unlike some of the satellites orbiting Jupiter or Saturn. Life could not have evolved under such conditions. Had Earth found itself too close to the sun, its atmosphere would have boiled off; it would have become a parched, barren inferno similar to Mercury. A slight adjustment outward—say, to where Venus is today—might have reduced solar radiation and spared Earth's atmosphere. But temperatures still would have been too high for water to condense. With no oceans to act as a dissolving agent, tremendous amounts of carbon dioxide would have remained trapped in the primordial atmosphere, creating a massive greenhouse effect and turning the planet into a furnace that would have killed off any hint of life.

Meandering dry riverbeds are all that remain of the water that may have coursed over the surface of Mars some 4 billion years ago (below). Rivers of red-hot lava (above) ploughed their way through the surface of Venus.

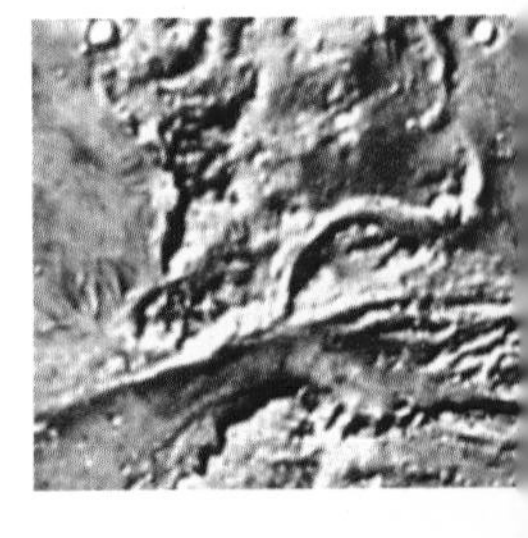

Such was, in fact, the fate of Venus. Its surface temperature is about five times higher than that of boiling water, above the melting point of lead. Of the nine planets in our solar system, Earth and Earth alone lies at just the right distance from the sun to be suitable for life.

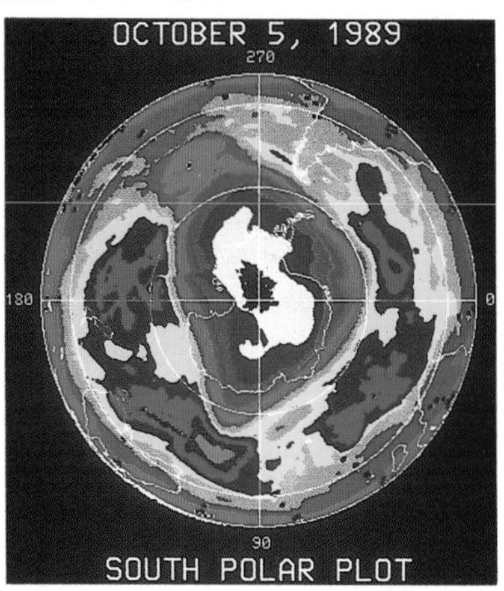

Some scientists believe that humans and their technology have begun to throw the planet's biosphere seriously out of kilter. A hole in the ozone layer, which shields us from the sun's harmful ultraviolet rays, recently opened up over Antarctica (above) and seems to be getting bigger. The reason, scientists suspect, is that ozone breaks down when it interacts with chlorofluorocarbons (CFCs), complex molecules used as aerosol propellants and refrigerants. Unless measures are taken to bring CFC production to a halt, the incidence of skin cancer will skyrocket and life on earth may never be the same.

The Search for Life on Other Planets

Are we to assume, then, there is no one in the entire universe except us humans? That hardly seems possible. After all, the observable universe is home to 100 billion galaxies, and every one of them harbors 100 billion stars. If, like our sun, each of those stars were the hub around which ten or so planets revolved, theoretically there could be as many as 100 thousand billion billion (10^{23}) planets in the universe. Since there is nothing unique about our position in space and time, why should we assume that life is unique to earth?

For today's astronomers, the search for extraterrestrial intelligence is no longer the stuff of science fiction. It is a serious, ongoing endeavor. But where should we look, and how? One way would be to send message-bearing probes into interstellar space, like bottled messages a castaway might toss into the sea.

The first two spacecraft ever designed to travel beyond the solar system, *Pioneer 10* (1972) and *11* (1973), each bear an aluminum plaque with drawings of a man and a woman and a schematic diagram indicating earth's

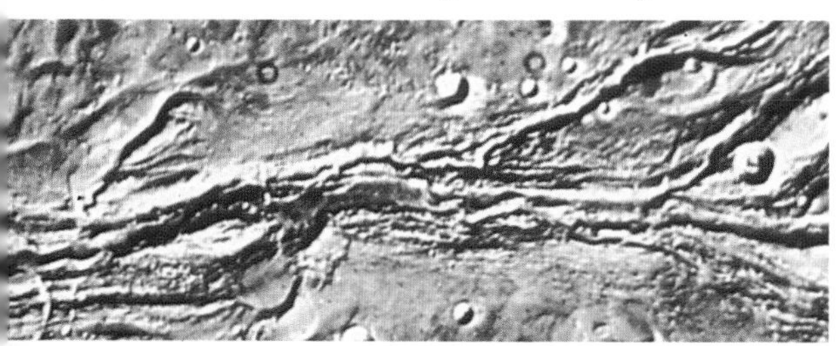

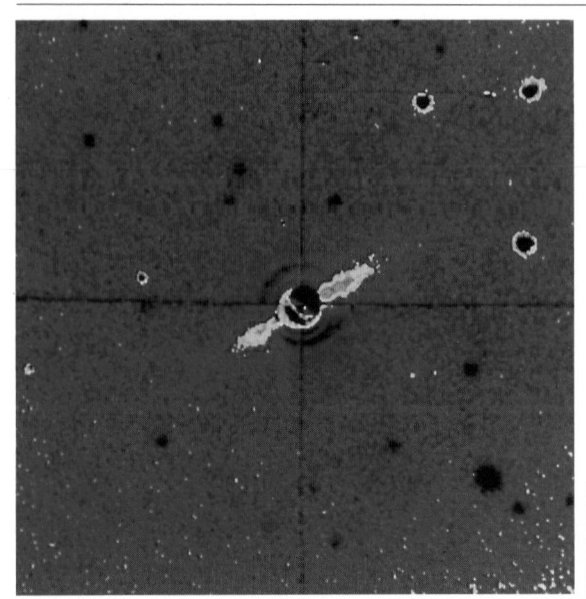

Earth-based telescopes have thus far been unable to observe planets outside the solar system directly, and the optical flaw in Hubble Space Telescope's primary mirror has dashed hopes of detecting any for the time being. Surely ours cannot be the only solar system in the Milky Way. Astronomers think they have spotted a solar system in the making around the star Beta Pictoris (infrared image, left). The light of the parent star has been blocked by a dark circular shield (center) to make the disk, an intense source of infrared radiation, easier to detect. Seen here edge on, the disk resembles the one that aggregated into our solar system but is much younger and may have formed as recently as a few hundred million years ago. Will one of the planets due to coalesce around Beta Pictoris eventually harbor life?

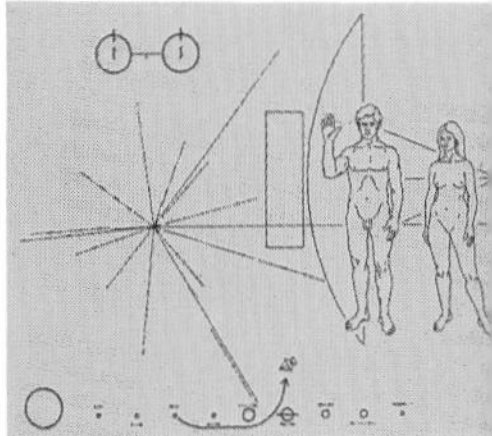

position in the galaxy to guide any aliens who might wish to call on us. The next two interstellar probes, *Voyager 1* and *2,* carry a video disk filled with images of life on earth and a copper phonograph record of the sounds of earth, from a Beethoven symphony to a jazz selection to the sound of a human kiss.

But sending message-bearing probes into space is hardly an ideal strategy for making contact with alien beings. For one thing, the odds of the probes being intercepted are low. For another, they are making their way across the endless reaches of the cosmos at a snail's pace. Although they are already moving much faster than anything on earth could, it will still take them some 80,000 years to reach the nearest star.

"Earth to Cosmos"

A far more effective strategy for making contact with extraterrestrials would be to beam out radio signals—or else listen in for them. Such messages would travel at the speed of light, the fastest possible speed in the universe. Radio transmissions would take only 4 years, not 40,000, to span the distance separating us from the nearest star. But where among the countless planets, stars, and galaxies in the universe would one send out or listen for signals? At which frequencies?

The first and to date only earth message was transmitted in 1974 by the world's largest radio telescope at Arecibo, Puerto Rico. To target as many potential listeners as they could, scientists aimed the signal at M13 in Hercules, a globular cluster of 300,000 gravitationally bound stars. The "channel" was the frequency at which hydrogen naturally radiates energy: Since hydrogen is by far the most abundant element in the universe, extraterrestrials, scientists reasoned, would be as familiar with its frequency as we are.

Even if an extraterrestrial never intercepts it, the phonograph record aboard *Voyager* (below) has already awakened billions of earthlings to the possibility of intelligent life beyond our planet.

The plaque attached to *Pioneer 10* and *11* (opposite right) prompted various sorts of comment and controversy. Some feminists criticized the decision to raise the male earthling's hand in greeting but not the female's. To have done so, the plaque's designers countered, might mislead aliens into thinking that one hand held in the air is a typical human posture. Another issue raised at the time was the female's external genitalia —omitted, according to codesigner Carl Sagan, in anticipation of NASA's disapproval.

The messages, all in binary code, include numbers from one to ten, the atomic numbers of a few basic elements, the formulas for the DNA molecule, and a schematic diagram of the solar system. As you read this, the Arecibo Interstellar Message is speeding toward the targeted globular star cluster. It is not due to arrive for another 24,000 years. When it finally does, any radio telescope in M13 pointed toward the sun should detect a millionfold increase in radio-wave intensity lasting three minutes.

Even if other beings intercept the message, understand it, and dash off a reply, we are still faced with the prospect of a 48,000-year wait.

Cosmic Eavesdropping on Aliens

But what if, instead of beaming messages out, we tried to tune them in? For all we know, transmissions from other civilizations may be surging through space this very instant. In October 1992, to commemorate the 500th anniversary of the discovery of America by Christopher Columbus, NASA initiated a twofold radio search project—a targeted search of sunlike stars at two specific frequencies and a comprehensive sky survey covering millions of channels—

Weightlessness is not expected to be a problem for 21st-century earthlings aboard space station *Freedom,* the future launching pad for expeditions to Mars and beyond.

in hopes of picking up artificial signals beamed by a distant civilization.

One day the unsettling silence of the universe may be broken at long last. It will mark one of the great turning points in the history of humankind.

Even if we never succeed in deciphering a message from another world, the event itself will have an incalculable impact. The certain knowledge that we are not alone will give us fresh insight into how our species fits into the scheme of things. The universe will seem a less hostile place, for we shall have learned that it is home to other beings with the capacity to marvel at the beauty of the cosmos.

Set in a natural hollow in Puerto Rico, the 1000-foot-diameter Arecibo radio telescope is the world's biggest. Because it is not steerable, its observational range is limited to celestial objects that the earth's rotation brings overhead into its "line of sight" near the zenith. In 1974, the Arecibo telescope transmitted a message at a frequency of 2380 MHz. A million times stronger than solar radiation at the same frequency, this signal theoretically could be picked up by extraterrestrial telescopes of comparable strength several thousand light-years away.

Beyond the Big Bang

Scientists have assembled tremendous amounts of data about the creation and composition of the universe. But there is still so much to be learned. What will the future bring?

DOCUMENTS

The universe baffles
scientists, inspires poets, and fills
all human beings with awe.

Cosmic Poetry

The universe has long been a source of inspiration for writers. In The Little Prince *(1943), French author and aviator Antoine de Saint-Exupéry (1900–44) uses the point of view of a human narrator journeying through the cosmos to expose our foibles and follies, cast the phenomena around us in a poetic light, and impart an occasional flash of insight: "What is essential is invisible to the eye," we learn from the fox in Saint-Exupéry's fable. Astrophysicists have since discovered that over 90 percent of the mass of the universe may indeed be invisible!*

In this excerpt Saint-Exupéry pokes fun at a habit astronomers have of referring to celestial objects by sets of letters and numbers instead of names.

I had thus learned a…fact of great importance: this was that the planet the little prince came from was scarcely any larger than a house!

But that did not really surprise me much. I knew very well that in addition to the great planets—such as the Earth, Jupiter, Mars, Venus—to which we have given names, there are also hundreds of others, some of which are so small that one has a hard time seeing them through the telescope. When an astronomer discovers one of these he does not give it a name, but only a number. He might call it, for example, "Asteroid 325."

I have serious reason to believe that the planet from which the little prince came is the asteroid known as B-612.

This asteroid has only once been seen through the telescope. That was by a Turkish astronomer, in 1909.

On making his discovery, the astronomer had presented it to the

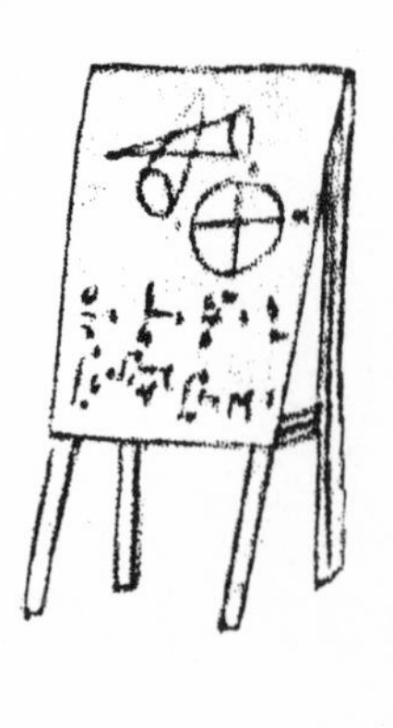

International Astronomical Congress, in a great demonstration. But he was in Turkish costume, and so nobody would believe what he said.

Grown-ups are like that…

Fortunately, however, for the reputation of Asteroid B-612, a Turkish dictator made a law that his subjects, under pain of death, should change to European costume. So in 1920 the astronomer gave his demonstration all over again, dressed with impressive style and elegance. And this time everybody accepted his report.

If I have told you these details about the asteroid, and made a note of its number for you, it is on account of the grown-ups and their ways. Grown-ups love figures. When you tell them that you have made a new friend, they never ask you any questions about essential matters. They never say to you, "What does his voice sound like? What games does he love best? Does he collect butterflies?" Instead, they demand: "How old is he? How many brothers has he? How much does he weigh? How much money does his father make?" Only from these figures do they think they have learned anything about him.

If you were to say to the grown-ups: "I saw a beautiful house made of rosy brick, with geraniums in the windows and doves on the roof," they would not be able to get any idea of that house at all. You would have to say to them: "I saw a house that cost $20,000." Then they would exclaim: "Oh, what a pretty house that is!"

Just so, you might say to them: "The proof that the little prince existed is that he was charming, that he laughed, and that he was looking for a sheep. If anybody wants a sheep, that is a proof that he exists." And what good would it do to tell them that? They would shrug their shoulders, and treat you like a child. But if you said to them: "The planet he came from is Asteroid B-612," then they would be convinced, and leave you in peace from their questions.

They are like that. One must not hold it against them. Children should always show great forbearance toward grown-up people.…

Human Greed

The fourth planet belonged to a businessman. This man was so much occupied that he did not even raise his head at the little prince's arrival.

"Good morning," the little prince said to him. "Your cigarette has gone out."

"Three and two make five. Five and

seven make twelve. Twelve and three make fifteen. Good morning. Fifteen and seven make twenty-two. Twenty-two and six make twenty-eight. I haven't time to light it again. Twenty-six and five make thirty-one. Phew! Then that makes five-hundred-and-one million, six-hundred-twenty-two thousand, seven-hundred-thirty-one."

"Five hundred million what?" asked the little prince.

"Eh? Are you still there? Five-hundred-and-one million—I can't stop...I have so much to do! I am concerned with matters of consequence. I don't amuse myself with balderdash. Two and five make seven...."

"Five-hundred-and-one million what?" repeated the little prince, who never in his life had let go of a question once he had asked it.

The businessman raised his head.

"During the fifty-four years that I have inhabited this planet, I have been disturbed only three times. The first time was twenty-two years ago, when some giddy goose fell from goodness knows where. He made the most frightful noise that resounded all over the place, and I made four mistakes in my addition. The second time, eleven years ago, I was disturbed by an attack of rheumatism. I don't get enough exercise. I have no time for loafing. The third time—well, this is it! I was saying, then, five-hundred-and-one millions—"

"Millions of what?"

The businessman suddenly realized that there was no hope of being left in peace until he answered this question.

"Millions of those little objects," he said, "which one sometimes sees in the sky."

"Flies?"

"Oh, no. Little glittering objects."

"Bees?"

"Oh, no. Little golden objects that set lazy men to idle dreaming. As for me, I am concerned with matters of consequence. There is no time for idle dreaming in my life."

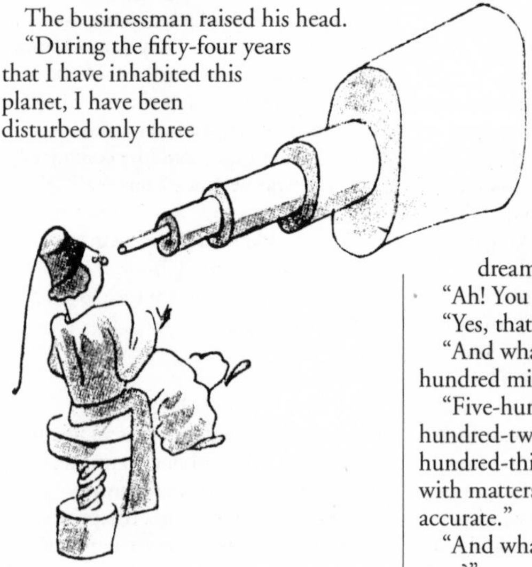

"Ah! You mean the stars?"

"Yes, that's it. The stars."

"And what do you do with five-hundred millions of stars?"

"Five-hundred-and-one million, six-hundred-twenty-two thousand, seven-hundred-thirty-one. I am concerned with matters of consequence: I am accurate."

"And what do you do with these stars?"

"What do I do with them?"

"Yes."

"Nothing. I own them."

"You own the stars?"

"Yes."

"But I have already seen a king who—"

"Kings do not *own,* they *reign over.* It is a very different matter."

"And what good does it do you to own the stars?"

"It does me the good of making me rich."

"And what good does it do you to be rich?"

"It makes it possible for me to buy more stars, if any are discovered."

"This man," the little prince said to himself, "reasons a little like my poor tippler. . . ."

Nevertheless, he still had some more questions.

"How is it possible for one to own the stars?"

"To whom do they belong?" the businessman retorted, peevishly.

"I don't know. To nobody."

"Then they belong to me, because I was the first person to think of it."

"Is that all that is necessary?"

"Certainly. When you find a diamond that belongs to nobody, it is yours. When you discover an island that belongs to nobody, it is yours. When you get an idea before any one else, you take out a patent on it: it is yours. So with me: I own the stars, because nobody else before me ever thought of owning them."

"Yes, that is true," said the little prince. "And what do you do with them?"

"I administer them," replied the businessman. "I count them and recount them. It is difficult. But I am a man who is naturally interested in matters of consequence."

The little prince was still not satisfied.

"If I owned a silk scarf," he said, "I could put it around my neck and take it away with me. If I owned a flower, I could pluck that flower and take it away with me. But you cannot pluck the stars from heaven. . . ."

"No. But I can put them in the bank."

"Whatever does that mean?"

"That means that I write the number of my stars on a little paper. And then I put this paper in a drawer and lock it with a key."

"And that is all?"

"That is enough," said the businessman.

"It is entertaining," thought the little prince. "It is rather poetic. But it is of no great consequence."

On matters of consequence, the little prince had ideas which were very different from those of the grown-ups.

"I myself own a flower," he continued his conversation with the businessman, "which I water every day. I own three volcanoes, which I clean out every week (for I also clean out the one that is extinct; one never knows). It is of some use to my volcanoes, and it is of some use to my flower, that I own them. But you are of no use to the stars.s . . ."

The businessman opened his mouth, but he found nothing to say in answer. And the little prince went away.

"The grown-ups are certainly altogether extraordinary," he said simply, talking to himself as he continued on his journey.

Antoine de Saint-Exupéry
The Little Prince
translated by Katherine Woods
1943

A Meaningless Universe?

The existence of human beings is implicated in every law of physics that governs the cosmos. The universe possesses just the properties needed to produce living creatures with consciousness and intelligence. That leaves us with a choice between two possible scenarios: Either it's all come about by chance, or the universe is governed by some underlying creative force.

"It's a question of using the right metaphor. *That's* why the inflationary universe is becoming so popular."

In his book Chance and Necessity, *Jacques Monod (1910–76) of France, recipient of the 1965 Nobel prize for physiology or medicine, comes down unequivocally on the side of chance and ends on a none-too-comforting note: We are "alone in the universe's unfeeling immensity."*

The initial elementary events which open the way to evolution in the intensely conservative systems called living beings are microscopic, fortuitous, and utterly without relation to whatever may be their effects upon teleonomic functioning.

But once incorporated in the DNA structure, the accident—essentially unpredictable because always singular—will be mechanically and faithfully replicated and translated: that is to say, both multiplied and transposed into millions or billions of copies. Drawn out of the realm of pure chance, the accident enters into that of necessity, of the most implacable certainties. For natural selection operates at the macroscopic level, the level of organisms.

Even today a good many distinguished minds seem unable to accept or even to understand that from a source of noise natural selection alone and unaided could have drawn all the music of the biosphere. In effect natural selection operates *upon* the products of chance and can feed nowhere else; but it operates in a domain of very demanding conditions, and from this domain chance is barred. It is not to chance but to these conditions that evolution owes its generally progressive course, its successive conquests, and the impression it gives of a smooth and steady unfolding....

When one ponders on the tremendous journey of evolution over the past three billion years or so, the prodigious wealth of structures it has engendered, and the extraordinarily effective teleonomic performances of living beings, from bacteria to man, one may well find oneself beginning to doubt again whether all this could conceivably be the product of an enormous lottery presided over by natural selection, blindly picking the rare winners from among numbers drawn at utter random.

While one's conviction may be restored by a detailed review of the accumulated modern evidence that this conception alone is compatible with the facts (notably with the molecular mechanisms of replication, mutation, and translation), it affords no synthetic, intuitive, and immediate grasp of the vast sweep of evolution. The miracle stands "explained"; it does not strike us as any less miraculous. As François Mauriac wrote, "What this professor says is far more incredible than what we poor Christians believe."...

The ancient covenant is in pieces; man knows at last that he is alone in the universe's unfeeling immensity, out of which he emerged only by chance. His destiny is nowhere spelled out, nor is his duty. The kingdom above or the darkness below: it is for him to choose.

Jacques Monod,
Chance and Necessity: An Essay on the Natural Philosophy of Modern Biology,
translated by Austryn Wainhouse, 1971

American physicist and Nobel laureate Steven Weinberg also argues for the meaninglessness of the universe and the pointlessness of human affairs.

A theory of the early universe has become so widely accepted that astronomers often call it "the standard model." It is more or less the same as what is sometimes called the "big bang" theory, but supplemented with a much more specific recipe for the contents of the universe....

In the beginning there was an explosion. Not an explosion like those familiar on earth, starting from a definite center and spreading out to engulf more and more of the circumambient air, but an explosion which occurred simultaneously everywhere, filling all space from the beginning, with every particle of matter rushing apart from every other particle. "All space" in this context may mean either all of an infinite universe, or all of a finite universe which curves back on itself like the surface of a sphere....

One type of particle that was present in large numbers is the electron, the negatively charged particle that flows through wires in electric currents and makes up the outer parts of all atoms and molecules in the present universe. Another type of particle that was abundant at early times is the positron, a positively charged particle with precisely the same mass as the electron.

In the present universe positrons are found only in high-energy laboratories, in some kinds of radioactivity, and in violent astronomical phenomena like cosmic rays and supernovas, but in the early universe the number of positrons was almost exactly equal to the number of electrons.

In addition to electrons and positrons, there were roughly similar numbers of various kinds of neutrinos, ghostly particles with no mass or electric charge whatever. Finally, the universe was filled with light. This does not have to be treated separately from the particles—the quantum theory tells us that light consists of particles of zero mass and zero electrical charge known as photons....

The universe will certainly go on expanding for a while. As to its fate after that, the standard model gives an equivocal prophecy: It all depends on whether the cosmic density is less or greater than a certain critical value....

If the cosmic density is *less* than the critical density, then the universe is of infinite extent and will go on expanding forever. Our descendants, if we have any then, will see thermonuclear reactions slowly come to an end in all the stars, leaving behind various sorts of cinder: black dwarf stars, neutron stars, perhaps black holes....

On the other hand, if the cosmic density is *greater* than the critical value, then the universe is finite and its expansion will eventually cease, giving way to an accelerating contraction.... The contraction is just the expansion run backward: after 50,000 million years the universe would have regained its present size, and after another 10,000 million years it would approach a singular state of infinite density....

Can we really carry this sad story all the way to its end, to a state of infinite temperature and density? Does time really have a stop some three minutes after the temperature reaches a thousand million degrees? Obviously,

"Is that *it?* Is that the *big bang?*"

we cannot be sure....

From these uncertainties some cosmologists derive a sort of hope. It may be that the universe will experience a kind of cosmic "bounce," and begin to reexpand.... But if the universe does reexpand, its expansion will again slow to a halt and be followed by another contraction, ending in another cosmic [expiration], followed by another bounce, and so on forever....

Some cosmologists are philosophically attracted to the oscillating model, especially because, like the steady-state model, it nicely avoids the problem of Genesis....

However all these problems may be resolved, and whichever cosmological model proves correct, there is not much of comfort in any of this. It is almost irresistible for humans to believe that we have some special relation to the universe, that human life is not just a more-or-less farcical outcome of a chain of accidents reaching back to the first three minutes, but that we were somehow built in from the beginning.

As I write this I happen to be in an airplane at 30,000 feet, flying over Wyoming en route home from San Francisco to Boston. Below, the earth looks very soft and comfortable—fluffy clouds here and there, snow turning pink as the sun sets, roads stretching straight across the country from one town to another. It is very hard to realize that this all is just a tiny part of an overwhelmingly hostile universe. It is even harder to realize that this present universe has evolved from an unspeakably unfamiliar early condition, and faces a future extinction of endless cold or intolerable heat. The more the universe seems comprehensible, the more it also seems pointless.

But if there is no solace in the fruits of our research, there is at least some consolation in the research itself. Men and women are not content to comfort themselves with tales of gods and giants, or to confine their thoughts to the daily affairs of life; they also build telescopes and satellites and accelerators, and sit at their desks for endless hours working out the meaning of the data they gather. The effort to understand the universe is one of the very few things that lifts human life a little above the level of farce, and gives it some of the grace of tragedy.

Steven Weinberg,
The First Three Minutes: A Modern View of the Origin of the Universe, 1988

American physicist Freeman Dyson takes issue with Monod's and Weinberg's viewpoint. As he sees it, "the universe in some sense must have known we were coming."

Professional scientists today live under a taboo against mixing science and religion....

Listen to...the biologist Jacques Monod: "Any mingling of knowledge with values is unlawful, forbidden," and the physicist Steven Weinberg: "The more the universe seems comprehensible, the more it also seems pointless."...

Monod and Weinberg, both of them first-rate scientists and leaders of research in their specialties, are expressing a point of view which does not take into account the subtleties and ambiguities of twentieth-century physics. The roots of their philosophical attitudes lie in the nineteenth century, not in the twentieth.

The taboo against mixing knowledge with values arose during the nineteenth century out of the great battle between the evolutionary biologists led by Thomas Huxley and the churchmen led by Bishop Wilberforce. Huxley won the battle, but a hundred years later Monod and Weinberg are still fighting the ghost of Bishop Wilberforce....

Looking back on the battle a century later, we can see that Darwin and Huxley were right. The discovery of the structure and function of DNA has made clear the nature of the hereditary variations upon which natural selection operates. The fact that DNA patterns remain stable for millions of years, but are still occasionally variable, is explained as a consequence of the laws of chemistry and physics.

There is no reason why natural selection operating on these patterns, in a species of bird that has acquired a taste for eating fish, should not produce a penguin's flipper. Chance variations, selected by the perpetual struggle to survive, can do the work of the designer. So far as the biologists are concerned, the argument from design is dead. They won their battle. But unfortunately, in the bitterness of their victory over their clerical opponents, they have made the meaninglessness of the universe into a new dogma. Monod states this dogma with his customary sharpness:

"The cornerstone of the scientific method is the postulate that nature is objective. In other words, the *systematic* denial that true knowledge can be got at by interpreting phenomena in terms of final causes, that is to say, of purpose."

Here is a definition of the scientific method that would exclude...some of the most lively areas of modern physics and cosmology.

It is easy to understand how some modern molecular biologists have come to accept a narrow definition of scientific knowledge. Their tremendous successes were achieved by reducing the complex behavior of living creatures to the simpler behavior of the molecules out of which the creatures are built. Their whole field of science is based on the reduction of the complex to the simple, reduction of the apparently purposeful movements of an organism to purely mechanical movements of its constituent parts.

To the molecular biologist, a cell is a chemical machine, and the protein and nucleic acid molecules that control its behavior are little bits of clockwork, existing in well-defined states and reacting to their environment by changing from one state to another. Every student of molecular biology learns his trade by playing with models built of plastic balls and pegs. These models are an indispensable tool for detailed study of the structure and function of nucleic acids and enzymes. They are, for practical purposes, a useful visualization of the molecules out of which we are built.

But from the point of view of a physicist, the models belong to the nineteenth century. Every physicist knows that atoms are not really little hard balls. While the molecular biologists were using these mechanical models to make their spectacular discoveries, physics was moving in a quite different direction.

For the biologists, every step down in size was a step toward increasingly simple and mechanical behavior. A bacterium is more mechanical than a frog, and a DNA molecule is more mechanical than a bacterium. But twentieth-century physics has shown that further reductions in size have an opposite effect. If we divide a DNA molecule into its component atoms, the atoms behave less mechanically than the molecule. If we divide an atom into nucleus and electrons, the electrons are less mechanical than the atom.

There is a famous experiment, originally suggested by Einstein, Podolsky and Rosen in 1935 as a thought experiment to illustrate the difficulties of quantum theory, which demonstrates that the notion of an electron existing in an objective state independent of the experimenter is untenable. The experiment has been done in various ways with various kinds of particles, and the results show clearly that the state of a particle has a meaning only when a precise procedure for observing the state is prescribed. Among physicists there are many different philosophical viewpoints, and many different ways of interpreting the role of the observer in the description of subatomic processes. But all physicists agree with the experimental facts which make it hopeless to look for a description independent of the mode of observation. When we are dealing with things as small as atoms and electrons, the observer or experimenter cannot be excluded from the description of nature. In this domain, Monod's dogma, "The cornerstone of the scientific method is the postulate that nature is objective," turns out to be untrue.

If we deny Monod's postulate, this does not mean that we deny the achievements of molecular biology.... We are not saying that chance and the mechanical rearrangement of molecules cannot turn ape into man. We are saying only that if as physicists we try to observe in the finest detail the behavior of a single molecule, the meaning of the words "chance" and "mechanical" will depend upon the way we make our observations.... "Chance" cannot be defined except as a measure of the observer's ignorance of the future. The laws leave a place for mind in the description of every molecule....

I think our consciousness is not just a passive epiphenomenon carried along by the chemical events in our brains, but is an active agent forcing the molecular complexes to make choices between one quantum state and another. In other words, mind is already inherent in every electron, and the processes of human consciousness differ only in degree but not in kind from the processes of choice between quantum states which we call "chance" when they are made by electrons.

Jacques Monod has a word for people who think as I do and for whom he reserves his deepest scorn. He calls us "animists," believers in spirits. "Animism," he says, "established a covenant between nature and man, a profound alliance outside of which seems to stretch only terrifying solitude. Must we break this tie because the postulate of objectivity requires it?" Monod answers yes: "The ancient covenant is in pieces; man knows at last that he is alone in the universe's unfeeling immensity, out of which he emerged only by chance."

I answer no. I believe in the covenant. It is true that we emerged in the universe by chance, but the idea of chance is itself only a cover for our ignorance. I do not feel like an alien in this universe. The more I examine the universe and study the details of its architecture, the more evidence I find that the universe in some sense must have known that we were coming.

Freeman Dyson
Disturbing the Universe
1979

"It is hard work to crank up a universe."

The Great Mysteries of Physics

"Why is the sky dark at night? Why does time go by?" The questions children ask have perplexed our greatest thinkers. The enigma of darkness at night was unraveled a few decades ago, thanks to Big Bang cosmology. But time is as baffling as ever.

Father of the modern detective story, Edgar Allan Poe (1809–49) also speculated about cosmological matters, including the paradox of darkness at night. Thought Poe, an infinite universe should contain an infinite number of stars forming a "wall of light," no matter which way an observer looks. One would therefore expect the sky to be as bright at night as it is during the day. His solution to the paradox is surprisingly close to what has since become the accepted scientific explanation: The night sky is dark because the universe is not eternal; and because space is so vast, the light of very distant stars has not had time to reach us.

No astronomical fallacy is more untenable, and none has been more pertinaciously adhered to, than that of the absolute *illimitation* of the Universe of Stars. The reasons for limitation, as I have already assigned them, *à priori,* seem to me unanswerable; but, not to speak of these, *observation* assures us that there is, in numerous directions around us, certainly, if not in all, a positive limit—or, at the very least, affords us no basis whatever for thinking otherwise. Were the succession of stars endless, then the background of the sky would present us an uniform luminosity, like that displayed by the Galaxy—*since there could be absolutely no point, in all that background, at which would not exist a star.* The only mode, therefore, in which, under such a state of affairs, we could comprehend the *voids* which our telescopes find in innumerable directions, would be by supposing the distance of the invisible background so immense that no ray

Edgar Allan Poe.

A starry night, in a 19th-century German engraving.

from it has yet been able to reach us at all. That this *may* be so, who shall venture to deny? I maintain, simply, that we have not even the shadow of a reason for believing that it *is* so....

We comprehend, then, the insulation of our Universe. We perceive the isolation of *that*—of *all* that which we grasp with the senses. We know that there exists one *cluster of clusters*—a collection around which, on all sides, extend the immeasurable wildernesses of a Space *to all human perception* untenanted. But *because* upon the confines of this Universe of Stars we are compelled to pause, through want of farther evidence from the senses, is it right to conclude that, in fact, there *is* no material point beyond that which we have thus been permitted to attain? Have we, or have we not, an analogical right to the inference that this perceptible Universe—that this cluster of clusters—is but one of a *series* of clusters of clusters, the rest of which are invisible through distance—through the diffusion of their light being so excessive, ere it reaches us, as not to produce upon our retinas a light-impression—or from there being no such emanation as light at all, in these unspeakably distant worlds—or, lastly, from the mere interval being so vast, that the electric tidings of their presence in Space, have not yet—through the lapsing myriads of years—been enabled to traverse that interval?

Have we any right to inferences—have we any ground whatever for visions such as these? If we have a right to them in *any* degree, we have a right to their infinite extension.

The human brain has obviously a leaning to the *"Infinite,"* and fondles the phantom of the idea. It seems to long with a passionate fervor for this impossible conception, with the hope of intellectually believing it when conceived. What is general among the whole race of Man, of course no individual of that race can be warranted in considering abnormal; nevertheless, there *may* be a class of superior intelligences, to whom the human bias alluded to may wear all the character of monomania.

My question, however, remains unanswered:—Have we any right to infer—let us say, rather, to imagine—an interminable succession of the "clusters of clusters," or of "Universes" more or less similar?

I reply that the "right," in a case such as this, depends absolutely upon the hardihood of that imagination which ventures to claim the right. Let me declare, only, that, as an individual, I myself feel impelled to the *fancy*—without daring to call it more—that there *does* exist a *limitless* succession of Universes, more or less similar to that of which *alone* we shall ever have cognizance—at the very least until the return of our own particular Universe into Unity. *If* such clusters of clusters exist, however—*and they do*—it is abundantly clear that, having had no part in our origin, they have no portion in our laws. They neither attract us, nor we them. Their material—their spirit is not ours—is not that which obtains in any part of our Universe. They could not impress our senses or our souls. Among them and us—considering all, for the moment, collectively—there are no influences in common. Each exists, apart and independently, *in the bosom of its proper and particular God.*

Edgar Allan Poe,
"Eureka: An Essay on the Material and Spiritual Universe," 1902,
originally published in 1848

We can freely move about in three-dimensional space (forward and backward, left and right, up and down); but our journey in time leads us inexorably forward, from cradle to grave. Irreversibility is also the rule on the macroscopic level, governed by the thermodynamic "arrow" that causes disorder to increase with time—a piece of ice melts in the sun, an abandoned cathedral will eventually collapse in a heap of debris, roses must fade. The third direction of time is the one in which the universe is expanding. The cosmological "arrow" exists because the galaxies are moving farther apart and the space between them getting bigger. British physicist Stephen Hawking tries to shed some light on the connections among the three "arrows" of time.

I shall discuss first the thermodynamic arrow of time. The second law of thermodynamics results from the fact that there are always many more disordered states than there are ordered ones. For example, consider the pieces of a jigsaw in a box. There is one, and only one, arrangement in which the pieces make a complete picture. On the other hand, there are a very large number of arrangements in which the pieces are disordered and don't make a picture....

Suppose the pieces of the jigsaw start off in a box in the ordered arrangement in which they form a picture. If you shake the box, the pieces will take up another arrangement. This will probably be a disordered arrangement in which the pieces don't form a proper picture, simply because there are so many more disordered arrangements. Some groups of pieces may still form parts of the picture, but the more you shake the box, the more likely it is that these groups will get broken up and the pieces will be in a completely jumbled state in which they don't form any sort of picture. So the disorder of the pieces will probably increase with time if the pieces obey the initial condition that they start off in a condition of high order....

Why do we observe that the thermodynamic and cosmological arrows point in the same direction? Or in other words, why does disorder increase in the same direction of time as that in which the universe expands?...

Conditions in the contracting phase would not be suitable for the existence of intelligent beings who could ask the question: Why is disorder increasing in the same direction of time as that in which the universe is expanding? The inflation in the early stages of the universe, which the no boundary proposal predicts, means that the universe must be expanding at very close to the critical rate at which it would just avoid recollapse, and so will not recollapse for a very long time. By then all the stars will have burned out and the protons and neutrons in them will probably have decayed into light particles and radiation. The universe would be in a state of almost complete disorder. There would be no strong thermodynamic arrow of time. Disorder couldn't increase much because the universe would be in a state of almost complete disorder already. However, a strong thermodynamic arrow is necessary for intelligent life to operate. In order to survive, human beings have to consume food, which is an ordered form of energy, and convert it into heat, which is a disordered form of energy. Thus intelligent life could not exist in the contracting phase of the universe. This is the explanation of why we observe that the thermodynamic and cosmological arrows of time point in the same direction.... To summarize, the laws of science do not distinguish between the forward and backward directions of time. However, there are at least three arrows of time that do distinguish the past from the future. They are the thermodynamic arrow, the direction of time in which disorder increases; the psychological arrow, the direction of time in which we remember the past and not the future; and the cosmological arrow, the direction of time in which the universe expands rather than contracts.

Stephen Hawking,
A Brief History of Time: From the Big Bang to Black Holes, 1988

Extraterrestrial Life

Perhaps more than any other astronomer, Carl Sagan has made the general public aware that intelligent alien beings may exist and that contact with a civilization beyond our solar system would have incalculable practical and philosophical repercussions.

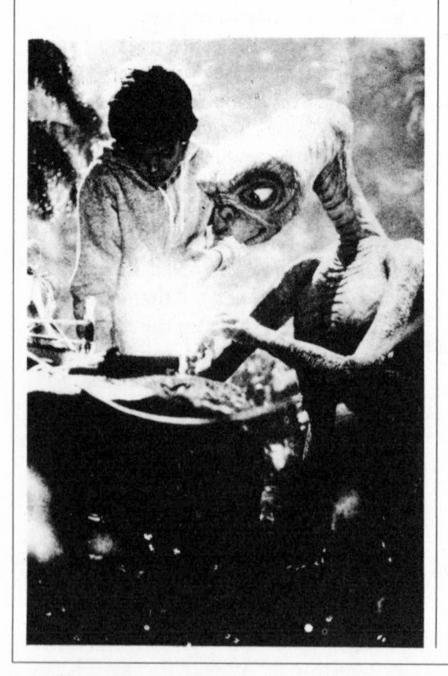

A Message from Earth

Carl Sagan recalls how he helped design the plaque now traveling through interstellar space aboard Pioneer 10 *and* 11.

Mankind's first serious attempt to communicate with extraterrestrial civilizations occurred on March 3, 1972, with the launching of the *Pioneer 10* spacecraft from Cape Kennedy. *Pioneer 10* was the first space vehicle designed to explore the environment of the planet Jupiter and, earlier in its voyage, the asteroids that lie between the orbits of Mars and Jupiter. Its orbit was not disturbed by an errant asteroid—the safety factor was estimated as 20 to 1. It approaches Jupiter in late December 1973, and then is accelerated by Jupiter's gravity to become the first man-made object to leave the Solar System. Its exit velocity is about 7 miles per second.

Pioneer 10 is the speediest object launched to date by mankind. But space is very empty, and the distances between the stars are vast. In the next 10 billion years, *Pioneer 10* will not enter the planetary system of any other star, even assuming that all the stars in the Galaxy have such planetary systems. The spacecraft will take about 80,000 years merely to travel the distance to the nearest star, about 4.3 light-years away.

But *Pioneer 10* is not directed to the vicinity of the nearest star. Instead, it will be traveling toward a point on the celestial sphere near the boundary of the constellations Taurus and Orion, where there are no nearby objects.

It is conceivable that the spacecraft will be encountered by an extra-

A scene from the film *E.T.*

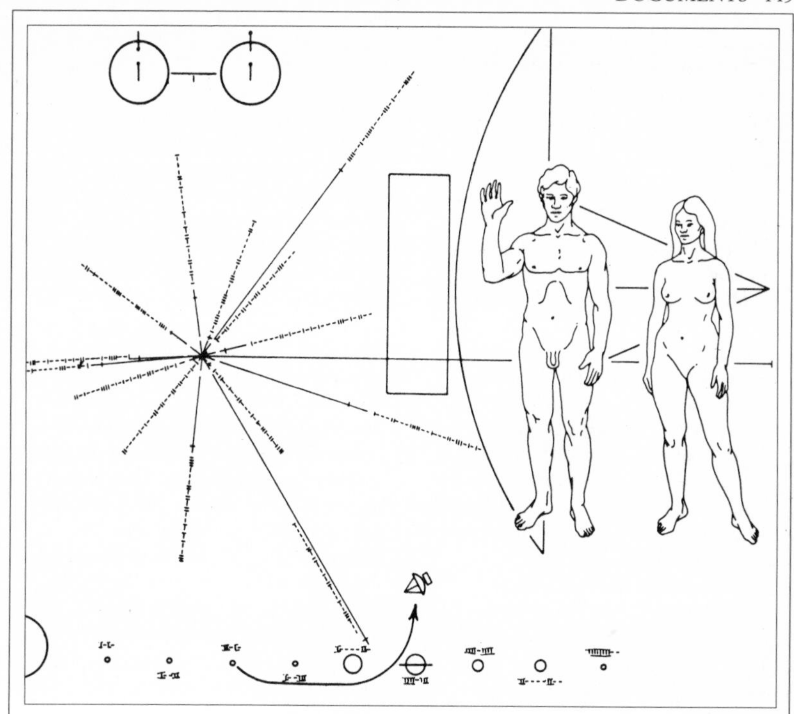

The plaque now traveling through space aboard *Pioneer 10* and *11.*

terrestrial civilization only if such a civilization has an extensive capability for interstellar space flight and is able to intercept and recover such silent space derelicts.

Placing a message aboard *Pioneer 10* is very much like a shipwrecked sailor casting a bottled message into the ocean—but the ocean of space is much vaster than any ocean on Earth.

When my attention was drawn to the possibility of placing a message in a space-age bottle, I contacted the *Pioneer 10* project office and NASA headquarters to see if there were any likelihood of implementing this suggestion. To my surprise and delight, the idea met with approval at all steps up the NASA hierarchy, despite the fact that it was—by ordinary standards—very late to make even tiny changes in the spacecraft. During a meeting of the American Astronomical Society in San Juan, Puerto Rico, in December 1971, I discussed privately various possible messages with my colleague Professor Frank Drake, also of Cornell. In a few hours we decided tentatively on the contents of the message. The human figures were added by my artist wife,

Linda Salzman Sagan. We do not think it is the optimum conceivable message for such a purpose: There were a total of only three weeks for the presentation of the idea, the design of the message, its approval by NASA, and the engraving of the final plaque. An identical plaque has been launched in 1973 on the *Pioneer 11* spacecraft, on a similar mission....

The message...is etched on a 6-inch by 9-inch gold-anodized aluminum plate, attached to the antenna support struts of *Pioneer 10.* The expected erosion rate in interstellar space is sufficiently small that this message should remain intact for hundreds of millions of years, and probably for a much longer period of time. It is, thus, the artifact of mankind with the longest expected lifetime.

The message itself intends to communicate the locale, epoch, and something of the nature of the builders of the spacecraft. It is written in the only language we share with the recipients: Science. At top left is a schematic representation of the hyperfine transition between parallel and antiparallel proton and electron spins of the neutral hydrogen atom. Beneath this representation is the binary number 1. Such transitions of hydrogen are accompanied by the emission of a radio-frequency photon of wavelength about 21 centimeters and frequency of about 1,420 Megahertz. Thus, there is a characteristic distance and a characteristic time associated with the transition. Since hydrogen is the most abundant atom in the Galaxy, and physics is the same throughout the Galaxy, we think there will be no difficulty for an advanced civilization to understand this part of the message. But as a check, on the right margin is the binary number 8 (1---) between two tote marks, indicating the height of the *Pioneer 10* spacecraft, schematically represented behind the man and the woman. A civilization that acquires the plaque will, of course, also acquire the spacecraft, and will be able to determine that the distance indicated is indeed close to 8 times 21 centimeters, thus confirming that the symbol at top left represents the hydrogen hyperfine transition.

Further binary numbers are shown in the radial pattern comprising the main part of the diagram at left center. These numbers, if written in decimal notation, would be ten digits long. They must represent either distances or times. If distances, they are on the order of several times 10^{11} centimeters, or a few dozen times the distance between the Earth and the Moon. It is highly unlikely that we would consider them useful to communicate. Because of the motion of objects within the Solar System, such distances vary in continuous and complex ways.

However, the corresponding times are on the order of 1/10 second to 1 second. These are the characteristic periods of the pulsars, natural and regular sources of cosmic radio emission; pulsars are rapidly rotating neutron stars produced in catastrophic stellar explosions. We believe that a scientifically sophisticated civilization will have no difficulty understanding the radial burst pattern as the positions and periods of 14 pulsars with respect to the Solar System of launch.

But pulsars are cosmic clocks that are running down at largely known rates. The recipients of the message must ask themselves not only where it was ever possible to see 14 pulsars arrayed in such a relative position, but also *when* it was possible to see them. The answers are: Only from a very small volume of the Milky Way Galaxy and in a single year in the history of the Galaxy. Within that small volume there are perhaps a thousand stars; only one is anticipated to have the array of planets with relative distances as indicated at the bottom of the diagram. The rough sizes of the planets and the rings of Saturn are also schematically shown. A schematic representation of the initial trajectory of the spacecraft launched from Earth and passing by Jupiter is also displayed. Thus, the message specifies one star in about 250 billion and one year (1970) in about 10 billion.

The content of the message to this point should be clear to an advanced extraterrestrial civilization, which will, of course, have the entire *Pioneer 10* spacecraft to examine as well. The message is probably less clear to the man on the street, if the street is on the planet Earth. (However, scientific communities on Earth have had little difficulty decoding the message.) The opposite is the case with the representations of human beings to the right. Extraterrestrial beings, which are the product of 4.5 billion years or more of independent biological evolution, may not at all resemble humans, nor may the perspective and line-drawing conventions be the same there as here. The human beings are the most mysterious part of the message.

Carl Sagan,
The Cosmic Connection: An Extraterrestrial Perspective, 1973

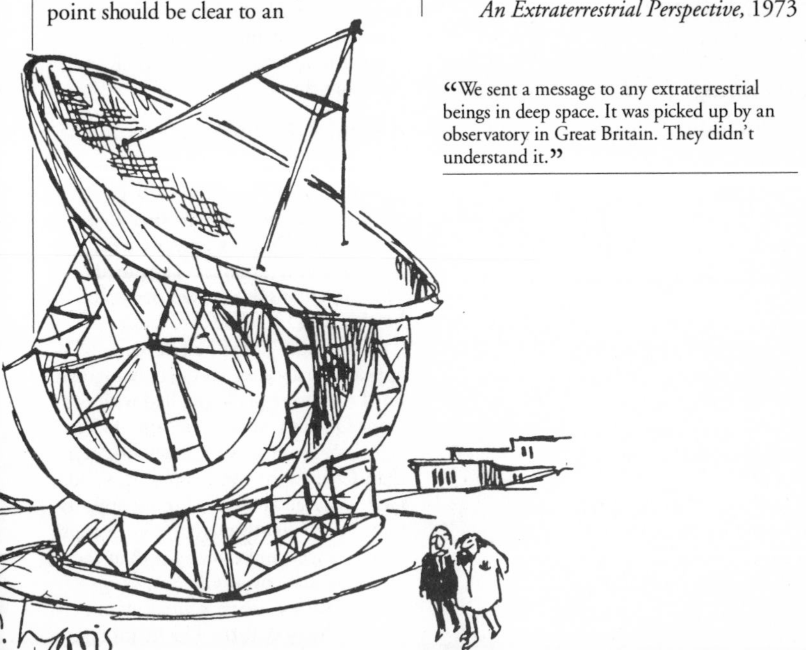

"We sent a message to any extraterrestrial beings in deep space. It was picked up by an observatory in Great Britain. They didn't understand it."

Hubble Opens New Vistas

The Hubble Space Telescope (HST) was to represent the culmination of decades of careful planning, fund-raising, and dreaming. Would it all come to nothing? True, the discovery of a major optical flaw was devastating, but—as this excerpt from a 1991 issue of Astronomy *magazine explains—the same technology that enabled HST to be conceived is helping to overcome its handicap. HST may yet help astronomers "unlock a secret of the universe."*

This past summer and fall [1990], mission planners put the two imaging instruments on HST—the Wide Field/Planetary Camera, or WF/PC (pronounced wiff-pick) for short, and the Faint Object Camera—through their paces. The resulting series of images proved that HST can do forefront astronomical research, most of it impossible to do from the ground, despite the flawed mirror....

Some observations remain largely unscathed by the spherical aberration and will be carried out with few or no changes. Other proposals are so severely affected that they won't be viable until HST is repaired. But a large percentage of the planned observations falls somewhere in between the two extremes—they can still be performed with the telescope in its aberrated state, but they require some type of modification, such as longer exposure times....

A View of Distant Neighbors

A long-anticipated target for Hubble was the planet Pluto and its large satellite, Charon. And the members of the Faint Object Camera science team didn't disappoint. They made Pluto the first solar system object to be observed by HST....

Pluto and Charon had never appeared truly distinct on a photograph until they were photographed with the Faint Object Camera [see page 115 right].... Pluto appears much brighter than Charon for two reasons: it is about twice as large as Charon and it appears to be covered with methane ice or snow, which reflects more sunlight than the water ice that covers Charon....

The WF/PC team went after an even bigger prize—Saturn. The first image taken of Saturn showed details never before captured on ground-based photographs, though such details have been seen by trained observers at the eyepiece of a telescope. A closeup of the north polar region of the planet shows atmospheric features, including an unusual hexagon-shaped structure. The details visible in the HST photos of Saturn are what you would see with the naked eye if Saturn were twice as far away as the Moon (which would make the planet a behemoth, measuring over 7 degrees across!).

Saturn also was the subject for the first "target of opportunity" of the mission. Although Hubble's observing schedule is normally fixed weeks, if not months, in advance, the telescope can target objects of special interest on fairly short notice.

Such a situation arose in late September [1990], when a large white spot appeared on Saturn's surface. This was the first large-scale atmospheric eruption on the planet in the past 60 years, and astronomers wanted to observe it with the sharp eye of Hubble. They got that opportunity in early November. The HST images allowed researchers to study the motions and vertical growth of the cloud in unprecedented detail, a key step to understanding what triggers these atmospheric events....

Fixing the Fuzziness

Image processing has played, and will continue to play, a big role in images from HST's cameras. The goal in most instances is to end up with an image that is as close to what the telescope would have given you if it were perfect. In some cases, that isn't quite as hard as it at first might seem.

For example, if you photograph a star with HST, you'll end up with an image that shows a bright core surrounded by a large, diffuse halo—the classic shape expected from a mirror with spherical aberration. If you can mathematically describe the shape of the flawed image produced by the telescope when it looks at a star, you can invert the process and mathematically describe what the star must have looked like given the shape of the flawed image. This technique is called deconvolution, and it works beautifully in some instances....

What Does the Future Hold?

The biggest loss for now...is the effort to measure the Hubble constant, the rate at which the universe is expanding. This was one of the three key projects scheduled for HST, but it now has to wait until the new Wide Field/ Planetary Camera, WF/PC II, is installed by shuttle astronauts in 1993. Hubble needs to observe and measure the brightnesses of 24th-magnitude and fainter stars in distant galaxies to determine the Hubble constant, and that's just about the point where the spherical aberration makes measuring stellar brightnesses nearly impossible....

During the winter months, the Scientific Assessment Program for the photometer and spectrometers will continue and some scientific projects will get under way. Other astronomers will be re-evaluating their proposals... to see whether and how they can make them work. And throughout it all, HST will keep orbiting Earth and doing as it's told, sending down data to astronomers hoping to unlock a secret of the universe.

Richard Talcott, *Astronomy,*
vol. 19, no. 2, February 1991

Glossary

active galaxies
Galaxies that emit most of their energy from a very compact central source a few light-hours to a few light-months in diameter, or billions of times smaller than the entire galaxy. Black holes millions of times more massive than the sun, capturing and consuming nearby stars, are thought to be the powerhouses of active galaxies

Andromeda Galaxy (M31)
Companion of the Milky Way Galaxy, 2.3 million light-years away. Together the two galaxies comprise the bulk of the mass of the Local Group

antimatter
Matter consisting of antiparticles, with the same properties as ordinary matter but the opposite electrical charge. The electron's antiparticle is the positron; the proton's, the antiproton. When particles and antiparticles come into contact, they annihilate each other and release photons. We live in a matter-dominated universe. Antimatter is extremely rare and observed only in cosmic-ray showers and high-energy particle accelerators

asteroids
Rocky, irregularly shaped minor planets in orbit around the sun between the orbits of Mars and Jupiter. They can measure up to about 600 miles in diameter

atom
Building block of matter consisting of a nucleus, made of protons and neutrons, and electrons in orbit around the nucleus. The nucleus contains 99.98 percent of the atom's mass, and its diameter (10^{-13} cm) is only one hundred-thousandth of the diameter of the atom (10^{-8} cm)

Big Bang theory
According to this theory, all the matter and energy in the universe was concentrated in an immensely hot and dense point that exploded 10 to 20 billion years ago. The universe has been expanding and cooling ever since

black dwarf
Invisible remnant of a white dwarf that has radiated all its energy away into space

black hole
Region of space produced when, for example, a star of more than five solar masses collapses, creating a gravitational field so powerful that matter and light cannot escape

blue giant
Young, massive (about 30 solar masses), superluminous (100,000 times the sun's luminosity) star

cannibalism
Process by which the gravitational forces of a galaxy capture less-massive galaxies in close proximity and send them spiraling in toward it. The less-massive galaxy is consumed and loses its identity as its stars mingle with those of the "cannibal" galaxy

comet
A body of ice and dust with a nucleus about 10 miles in diameter. Comets are visible only when they travel close enough to the sun to reflect its light. As solar heating evaporates the ice on the nucleus surface, a comet develops a tail hundreds of millions of miles long that always points away from the sun

cosmic background radiation
Microwave radiation that pervades the entire universe, believed to be the "afterglow" of the period some 300,000 years after the Big Bang. The cosmic background radiation and the expansion of the universe provide crucial supporting evidence for the Big Bang scenario

cosmic rays
Extremely high-energy particles, mostly protons and electrons

cosmology
Study of the origin, evolution, and structure of the universe

dark matter
Invisible matter deduced by observing the motion of stars, gas, and galaxies and the relative abundance of chemical elements produced in the Big Bang. The "missing mass" may make up between 90 and 98 percent of the total mass of the universe

density variations
Fluctuations in the density of the universe, detected as minute temperature differences in the cosmic background radiation and thought to be the seeds of protogalaxies

dwarf galaxy
Low-mass galaxy six times smaller (average diameter 15,000 light-years) and one thousand to ten thousand times less massive than a normal galaxy. May be elliptical or irregular; spiral dwarf galaxies have not been observed

electromagnetic force
Fundamental force that operates only between charged particles and causes oppositely charged particles to attract and particles of like charge to repel. Electromagnetic force binds electrons to nuclei to form atoms and binds atoms together to form solid matter

electromagnetic spectrum
Entire family of radiation, including, in order of decreasing energy: gamma rays, X rays, ultraviolet radiation, visible light, infrared radiation, microwaves, and radio waves

electron
Subatomic particle with a negative charge

elementary particles
New discoveries in subatomic physics have made it necessary to redefine the basic building blocks of matter and radiation. Electrons, neutrinos, photons, and quarks are considered elementary particles

elliptical galaxy
Galaxy observed as an oval-shaped system, generally composed of old stars and containing little or no gas and dust

galactic disk
Flattened aggregation of stars, gas, and dust in a spiral galaxy. The average disk is some 90,000 light-years in diameter and 300 light-years thick

galactic halo
Spherical region around a spiral galaxy populated by old stars and globular clusters. Observations suggest that it is surrounded by an invisible halo some ten times bigger and more massive than the visible one

galaxy
System of stars (10 million in a dwarf galaxy, 10 trillion in a giant galaxy) held together by gravitation. Galaxies are the fundamental units of the cosmic macrostructure

galaxy cluster
Grouping of thousands of galaxies held together by gravity

galaxy group
Collection of about twenty galaxies held together by gravity, some 6 million light-years across and averaging between 1 and 10 trillion solar masses

gamma rays
Most energetic form of electromagnetic radiation

globular cluster
Compact, spherical aggregation of about 100,000 old stars held together by gravity

gravity
Force responsible for attraction between all matter

heavy elements
All chemical elements with nuclei heavier than helium. Also referred to as "metals," heavy elements are built up by nuclear fusion in the interiors of stars

helium
Element with a nucleus of two protons and two neutrons. A second, far less abundant isotope has two protons and one neutron. Produced during the first 3 minutes of the universe, helium comprises about 25 percent of the mass of the universe

hydrogen
Lightest of all elements, consisting of one proton and one electron. Hydrogen atoms make up 75 percent of the universe by mass

interstellar dust
Solid microscopic particles of matter that condense in the distended outer layers of red giant stars. They absorb the blue light of stars, causing them to dim and redden

irregular galaxy
Galaxy that is neither spiral nor elliptical. Often they are dwarf galaxies and contain many young stars and large amounts of gas and dust

Kelvin
The Kelvin temperature scale measures the motion of atoms. The coldest possible temperature, called "absolute zero," where there is no motion, is 0 K or –273 Celsius. Water freezes at 273 K and boils at 373 K

light-year
The distance light travels in a vacuum in one year. One light-year is equal to 6×10^{12} miles

Local Group
Grouping of galaxies—extending over a region of space about 3 million light-years in diameter—of which the Milky Way and Andromeda are the principal and most massive members (1 trillion solar masses each). It also includes dwarf galaxies ranging from 10 million to 10 billion solar masses

Local Supercluster
Huge flattened supercluster of galaxies that contains the Local Group. It is also known as the Virgo Supercluster

M (Messier)
In a catalogue, the 18th-century French astronomer Charles Messier listed roughly the one hundred brightest nonstellar objects in the sky, assigning each of them a number. Astronomers often refer to these objects by their "M" numbers. For example, the Crab Nebula (a supernova remnant), the first object in Messier's list, is called M1

meteor
Streak of light produced by frictional heating of cosmic debris made of rock or metal that enters the earth's atmosphere

meteorite
Fragment of cosmic debris that survives burning up in the earth's atmosphere and lands on the planet's surface

Milky Way
Figurative name for our galaxy

molecule
Stable grouping of two or more atoms bound together by electromagnetic force

nebula
Celestial object with a diffuse nonstellar appearance. Term used to designate galaxies or masses of interstellar gas and dust

NGC (New General Catalogue)
List of nearly 15,000 faint nonstellar objects. Astronomers often refer to galaxies and nebulas by their NGC numbers. For example, the supernova remnant called the Crab Nebula is NGC 1952

neutrino
Electrically neutral elementary particle governed only by the weak nuclear force (and, if it has mass, by gravitation). Produced in huge quantities in the first moments after the Big Bang and to a lesser extent in stellar cores and supernovas, neutrinos could account for the bulk of the mass of the universe if a million of them should be found to equal the mass of a single electron. If ten thousand of them should be found to equal the mass of an electron, their gravity could theoretically reverse the expansion of the universe. It is not yet known whether neutrinos have mass

neutron
Electrically neutral subatomic particle composed of three quarks. Like the proton, it is a constituent of atomic nuclei. Neutrons are 1838 times more massive than electrons

neutron star
Extremely compact and superdense star composed almost entirely of neutrons, created when a dying star (with mass between 1.4 and 5 times the mass of the sun) has exhausted its nuclear fuel and collapses under its own gravity. Neutron stars spin very rapidly and emit beams of radiation that whip past the earth at regular intervals. Since the beams are detected as energy pulses, neutron stars are commonly known as pulsars

nuclear fission (nucleosynthesis)
Joining of two atomic nuclei to form a heavier nucleus, under the influence of the strong nuclear force

nucleus
Core of an atom, consisting of protons and neutrons bound together by the strong nuclear force. It has a positive charge equal to the total number of proton charges

ozone
Molecules made of three oxygen atoms. The ozone layer high in the atmosphere shields the earth against ultraviolet radiation

photon
Particle of light whose energy determines its nature in the electromagnetic spectrum

planetary nebula
Glowing shell of gas ejected by a dying star (with mass less than 1.4 solar masses) as the star collapses under its own gravity

prominence
Hot, glowing gas rising above the solar surface

protogalaxy
Young galaxy in the process of forming

proton
Positively charged particle composed of three quarks. One of the two constituents of all atomic nuclei

protostar
Young star in the process of forming

pulsar
See neutron star

quark
Elementary particle and a building block of protons and neutrons. Quarks have electrical charges that are either one-third or two-thirds of the electron charge and are subject to the strong nuclear force. Quarks have yet to be observed

quasar
Quasars are the remotest and brightest objects in the universe. They are at the centers of galaxies, and their unbelievable energy is probably generated by black holes a billion times more massive than the sun that "feed" on stars from the host galaxy

radio astronomy
Branch of astronomy concerned with the study of celestial objects emitting radio waves

red giant
Star that burns helium because it has exhausted its central hydrogen fuel. The outward pressure produced by burning helium swells the star's outer layers to many times its initial size. At the same time, the star's surface layers cool, reddening its light

retrograde motion
Apparent reverse motion of a planet in relation to background stars

solar corona
Outermost part of the sun, millions of miles in extent, consisting of highly rarefied gas at temperatures of millions of degrees

solar wind
Stream of electrons, protons, and helium nuclei coming from the sun's surface

spiral galaxy
Flattened, disklike system of stars and interstellar gas and dust. Its pinwheel-like arms of bright, young stars wind outward in the plane of the disk from a central spherical cluster

star
Sphere of gas consisting of 98 percent hydrogen and helium and 2 percent heavy elements, created when the compressive force of gravitation is exactly counterbalanced by outward radiation pressure from the nuclear fusion reactions in its center

Steady State theory
Theory that maintains that the universe is unchanging over time. Its overall

appearance is the same wherever and whenever it is observed. Matter is continuously created to compensate for the empty space the observed expansion of the universe is creating between galaxies

stellar nursery
Cloud of interstellar gas and dust collapsing under its own gravity and fragmenting into hundreds of protostars

strong nuclear force
Strongest of the four fundamental forces. Binds quarks together to form protons and neutrons, and joins protons and neutrons to form atomic nuclei

sunspot
Area on the sun's surface that looks dark because its temperature (4000 K) is considerably lower than the average surface temperature (6000 K)

supercluster
Flattened aggregation of tens of thousands of galaxies held together by gravity

supernova
Energetic explosion signaling the death of a star more massive than 1.4 solar masses

thermonuclear reaction
Reaction resulting from the high-speed collision of atomic nuclei moving rapidly because they are at a high temperature

weak nuclear force
Fundamental force responsible for radioactive decay

white dwarf
Small celestial object (about the size of earth) created when a star of less than 1.4 solar masses exhausts its nuclear fuel and collapses under its own gravity

X rays
Most energetic form of electromagnetic radiation after gamma rays

Further Reading

Asimov, Isaac, *To the Ends of the Universe,* Walker and Co., New York, 1976

Davies, P. C. W., *Superforce: The Search for a Grand Unified Theory of Nature,* Simon and Schuster, New York, 1984

Davis, Joel, *Journey to the Center of Our Galaxy: A Voyage in Space and Time,* Contemporary Books, Chicago, 1991

Dyson, Freeman, *Disturbing the Universe,* Harper and Row, New York, 1979

Friedman, Herbert, *The Astronomer's Universe,* Norton, New York, 1990

Gribbin, John R., and Martin J. Rees, *Cosmic Coincidences: Dark Matter, Mankind, and Anthropic Cosmology,* Bantam Books, New York, 1989

Harrison, James A., ed., *The Complete Works of Edgar Allan Poe,* AMS Press, repr. of 1902 ed., New York, 1965

Hawking, Stephen W., *A Brief History of Time: From the Big Bang to Black Holes,* Bantam Books, New York, 1988

Hoyle, Fred, *The Intelligent Universe,* Holt, Rinehart and Winston, New York, 1983

Illingworth, Valerie, *The Facts on File Dictionary of Astronomy,* 2nd ed., New York, 1985

Lederman, Leon M., and David N. Schramm, *From Quarks to the Cosmos,* Scientific American Library, New York, 1989

Monod, Jacques, *Chance and Necessity: An Essay on the Natural Philosophy of Modern Biology,* trans. by Austryn Wainhouse, Knopf, New York, 1971

Reeves, Hubert, *Atoms of Silence: An Exploration of Cosmic Evolution,* trans. by Ruth A. Lewis and John S. Lewis, MIT Press, Cambridge, Massachusetts, 1984

Ronan, Colin A., *The Natural History of the Universe from the Big Bang to the End of Time,* Macmillan, New York, 1991

Rowan-Robinson, Michael, *Our Universe: An Armchair Guide,* W. H. Freeman and Co., New York, 1990

Sagan, Carl, *The Cosmic Connection: An Extraterrestrial Perspective,* Doubleday, New York, 1973

———, *Cosmos,* Ballantine Books, New York, 1985

———, *The Dragons of Eden: Speculations on the Evolution of Human Intelligence,* Random House, New York, 1977

Saint-Exupéry, Antoine de, *The Little Prince,* trans. by Katherine Woods, Harcourt Brace Jovanovich, New York, 1943

Talcott, Richard, "Hubble Opens New Vistas," *Astronomy,* vol. 19, no. 2, February 1991

Thuan, Trinh Xuan, *The Secret Melody,* Oxford University Press, New York, 1994

Weinberg, Steven, *The First Three Minutes: A Modern View of the Origin of the Universe,* Basic Books, New York, 1988

List of Illustrations

Index

Page numbers in italics refer to captions and/or illustrations

Acknowledgments

The author is grateful to NASA for the wealth of material it so kindly made available to him. The publishers would like to thank Emmanuel Calamy and Pierre-Marie Valat for their illustrations

Photograph Credits

All rights reserved 23, 38a, 59a, 59b, 66, 67, 103, 120b, 121. Anglo-Australian Telescope Board 34l, 50l, 78, 81b, 89c, 129. Archives Larousse, Paris 27a, 28a, 29b, 33l, 64a, 68a, 72l. Astronomical Society of the Pacific 42–3, 83b, 124l. ATT Archives, Warren 68–9b. Bibliothèque Nationale, Paris 13, 17, 19, 20, 21, 22a, 26, 120a. British Museum, London 16a, 46–7, 101a. Emmanuel Calamy 36, 48–9a, 72–3, 74, 75a, 76–7, 88–9a, 89b, 95b, 96, 97, 104–5b, 106. California Institute of Technology, Palomar Observatory, Pasadena 33r, 39a, 47r, 94b. CERN, Geneva 60. Ciel et Espace 46l, 127al. Ciel et Espace/Brunier 116b. Ciel et Espace/CFHT/Gregory 1. Ciel et Espace/ESO 6. Ciel et Espace/Hardy 125l. Ciel et Espace/JPL 110a, 110b, 110c. Ciel et Espace/NAIC-NSF 127ar. Ciel et Espace/Sisk 2, 8, front cover. Ciel et Espace/USGS 111a, 111b. Conservatoire National des Arts et Métiers, Paris 32. Dagli-Orti 14–5, 100–1b. DITE/IPS/NASA 63. Jean Effel 61. European Southern Observatory, Munich 34r, 35b, 48b, 92. Explorer, Paris 118a, 142. Explorer/Krafft 102, 117. Explorer/Plisson 119. Gallimard 130, 131, 132. Giraudon 27b. Sidney Harris, New York 136, 138, 141. McDonald Observatory 49b. Max Planck Institut, Bonn 40–1. Mount Stromlo and Siding Spring Observatories 44–5. NASA 11, 30, 31, 37a, 37bl, 37br, 44b, 44–5c, 52a, 52b, 54, 69a, 75b, 79, 82a, 82–3, 84a, 84b, 85, 86, 87a, 87b, 90, 91, 93, 94a, 95a, 104–5a, 107, 111a, 111c, 112al, 112ar, 112b, 113a, 113b, 114, 115a, 115b, 122a, 122–3b, 123a, 124r, 125r, 126–7, 147, back cover. NASA/NOAO 53, 80–1a. Nationalhistoriske Museum, Frederiksborg, Denmark 22b. Palais de la Découverte 118b. Réunion des Musées Nationaux 12, 14, 18, 25a, 62. Ronan Picture Library, Somerset 28b, 29a. Scala, Florence 24, 25b. Science Photo Library/George Fowler 3, 5, 9. Science Photo Library/Gordon 4. Science Photo Library/Lorre 7. Sipa Press 70–1. Smithsonian Institution, Washington, D.C. 35a, 55l, 58, 100a, front cover, spine. Trinh X. Thuan 50r, 51. Pierre-Marie Valat 38–9b, 56–7, 98–9, 108–9. Yerkes Observatory, Chicago 116a

Text Credits

Grateful acknowledgment is made for use of material from the following works: *Disturbing the Universe* by Freeman Dyson. Copyright © 1979 by Freeman J. Dyson. Reprinted by permission of HarperCollins Publishers Inc. (pp. 137–9); Harrison, James A., ed., *The Complete Works of Edgar Allan Poe,* Vol. XVI, AMS Press, Inc., (New York, 1965 reprint) (pp. 140–2); *A Brief History of Time* by Stephen W. Hawking. Copyright © 1988 by Stephen W. Hawking. Used by permission of Bantam Books, a division of Bantam Doubleday Dell Publishing Group, Inc. Reprinted by permission of Stephen Hawking and Writers House, Inc. (p. 143); *Chance and Necessity* by Jacques Monod, trans. A. Wainhouse. Copyright © 1971 by Alfred A. Knopf, Inc. Reprinted by permission of the publisher (pp. 134–5); *The Cosmic Connection* by Carl Sagan. Copyright © 1973 by Carl Sagan and Jerome Agel. Used by permission of Doubleday, a division of Bantam Doubleday Dell Publishing Group, Inc. (pp. 144–7); *Cosmos.* Copyright © 1980 Carl Sagan. All rights reserved. Reprinted by permission of the author (inside front cover); *The Little Prince* by Antoine de Saint-Exupéry, copyright 1943 and renewed 1971 by Harcourt Brace & Company, reprinted by permission of the publisher (pp. 130–3); "Hubble Opens New Vistas" by Richard Talcott (ASTRONOMY Magazine, Feb., 1991), reproduced by permission of the publisher. Copyright 1993 by Kalmbach Publishing Co. (pp. 148–9); *The First Three Minutes: A Modern View of the Universe* by Steven Weinberg. Copyright © 1977 by Steven Weinberg. Reprinted by permission of Basic Books, a division of HarperCollins Publishers Inc. (pp. 135–7)

A native of Hanoi, Vietnam, Trinh Xuan Thuan attended the Lycée Français in Saigon and studied at the California Institute of Technology and Princeton University. Since 1976 he has been a professor of astronomy at the University of Virginia, where he currently teaches a course in astronomy for poets. He specializes in extragalactic astronomy and has written many articles on galaxy formation and evolution. Professor Thuan has also published two other books, the widely acclaimed *The Secret Melody* (English translation, 1994) and *Un Astrophysicien* (1992).

Translated from the French by I. Mark Paris

Project Manager: Sharon AvRutick
Typographic Designer: Elissa Ichiyasu
Assistant Editor: Jennifer Stockman
Design Assistant: Penelope Hardy

Library of Congress Catalog Card Number: 93–70491

ISBN 0–8109–2815–9

Published in 1993 by Harry N. Abrams, Incorporated, New York
A Times Mirror Company

Printed and bound in Italy by Editoriale Libraria, Trieste